STATISTICS AND DYNAMICS OF URBAN POPULATIONS

Statistics and Dynamics of Urban Populations

Empirical Results and Theoretical Approaches

Marc Barthelemy and Vincent Verbavatz

OXFORD
UNIVERSITY PRESS

OXFORD
UNIVERSITY PRESS

Great Clarendon Street, Oxford, OX2 6DP,
United Kingdom

Oxford University Press is a department of the University of Oxford.
It furthers the University's objective of excellence in research, scholarship,
and education by publishing worldwide. Oxford is a registered trade mark of
Oxford University Press in the UK and in certain other countries

Published in the United States of America by Oxford University Press
198 Madison Avenue, New York, NY 10016, United States of America

British Library Cataloguing in Publication Data

Data available

Library of Congress Control Number: 2023945985

ISBN 9780192867544
DOI: 10.1093/oso/9780192867544.001.0001

Printed and bound by
CPI Group (UK) Ltd, Croydon, CR0 4YY

Links to third party websites are provided by Oxford in good faith and
for information only. Oxford disclaims any responsibility for the materials
contained in any third party website referenced in this work.

FSC
www.fsc.org

MIX
Paper | Supporting
responsible forestry
FSC® C013604

Preface

Although they appeared first almost ten thousand years ago, cities, which at the beginning of the twenty-first century hold more than half of the world's population, are one of the most successful and complex forms of social organization. Yet it is anything but natural to think that it is easy to provide food and goods to very large populations concentrated in small areas and that billions of people can afford to live far from the means of production on which they depend for their survival. Crises, natural disasters and wars show that once the machinery breaks down, it is better to avoid cities.

However, when everything is going well, cities shine because of the social synergies they create: emulation, increased productivity and economies of scale. The urban model seems to be one of efficiency and even sobriety. This is how history, geography, sociology and economics have understood cities since they became a subject of study.

But are these synergies quantifiable? If cities are more productive, more ecological or more efficient, by how much? These questions are of interest to both researchers and public authorities in the orientation of their choices. Asking them requires us to stop for a moment. Indeed, they presuppose that cities can be seen as objects which can be studied as a single concept; that Paris, Jakarta, Lagos, Des Moines, Pondicherry are somehow the same thing. Like any structuralist effort of conceptualization, this should not be considered trivial. What, then, authorizes us to pool together different realities under the same name?

The answer is essentially empirical. Cities surprise us with their similarities and their universality, despite the immensity and diversity of their populations. Everywhere, cities are made of buildings, streets, sewers. In every city there are (a lot of) workers and dwellers. And actually, even more complex phenomena seem to be universal: gentrification, segregation and public networks are more general than specific.

When there is universality, it is natural to look for common causes: this is the object of a "science of cities." The aim of this science is to establish quantitative laws, which are as simple as possible and can describe and account for what cities, as urban systems, have in common. All these relationships are based first and foremost on sound empirical analysis, made possible by the proliferation of data sources over the last 20 years. Without these data, no trustworthy modeling is possible.

This book takes place in the framework of this science of cities. As physicists, we have two fundamental guidelines that we will follow through this work: reality and parsimony. By the concept of reality, we mean that every result we will derive will be either empirical or consistent with empirical observations. In contrast, we think that theoretical results are of no avail if not compared with data.

By the concept of parsimony, we mean that we are looking for models that are as simple as possible. Any supplementary assumption, any new parameter, has a cost: it necessarily reduces the universality of its model. This need for a relatively small number of parameters is a modern formulation of Occam's razor: *pluralitas non est ponenda sine necessitate* [1]. A high number of parameters may indeed be a sign of redundancy, and thus evidence that a level of unification has been missed. Throughout this book, we will always focus on approaches that try to keep the number of parameters and assumptions as minimal as possible.

Together, reality and parsimony are what can make a model universal—that is, able to reproduce empirical facts in a wide range of diverse situations. If cities are universal, such models should exist.

Cities have multiple aspects. In this book we focus on a relatively simple feature, but a crucial one: their population. Indeed, the population is undoubtedly the most important urban variable: fairly reliable and easy to measure, even over long periods of time—it has been possible in Europe to obtain very good reliable data per household since the fourteenth century—, the population is also a first-order indicator of many urban characteristics.

In this respect, it is notable that modern cities are not all of the same size, a seemingly trivial fact which disqualifies the idea of an "optimal size" for cities. There is no condition that should lead all the cities of the world to converge toward an equilibrium—typically between increasing and decreasing returns—implying a single size. On the contrary, the most recognized and characterized urban fact so far is the idea of a regular and universal *inequality of sizes* among cities, long approximated by Zipf's law (Zipf, 1949) that was first discovered for cities by the physicist Felix Auerbach (Auerbach, 1913).

Zipf's law characterizes the hierarchical organization of cities, and implies that in any country, the city with the largest population is generally twice as large as the next largest, and so on. It is a signature of the very large heterogeneity of city sizes, what Krugman (1996) called the "mystery of urban hierarchy". Zipf's law is a *static* result, showing how urban population is distributed at equilibrium. As we will show, it has been disproved in recent years.

Another important aspect of urban population is its *dynamical* behavior: how cities grow and their rank evolves in time. Typical models of urban growth are Gibrat's model (Gibrat, 1931), Simon's model (Simon, 1955) and Gabaix's model (Gabaix, 1999). We will elaborate on those later. Notably, we will present our own model of city growth (Verbavatz and Barthelemy, 2020). This model will rely on advanced mathematical objects that will be introduced throughout this text.

This book is divided into four main parts. The first part discusses population and its importance for characterizing cities. In Chapter 1, we will see what we mean by urban population and the limitations that are brought to this measure by the problem of *defining cities*. In Chapter 2, we will explain why the population matters, i.e. how and why it is a good proxy for other urban variables. In particular, we will elaborate on the concept of *scaling laws* in cities and the limits to this reasoning.

The second part of the book is about how cities can be ranked and compared with each other. In Chapter 3, we will introduce the cornerstone of quantitative urban studies, *Zipf's law*, which compares cities at equilibrium. For non-experts, we will

[1]Plurality should not be assumed without necessity.

present Zipf's law amid more general results on power-laws and how to characterize them in empirical data. This more theoretical discussion will show us that empirical arguments in favor of Zipf's law are actually weak and that Zipf's law can be easily refuted as a spurious observation. Conversely, in Chapter 4, we will look at how cities rank in a dynamical manner. The discussion will present some fundamental results of the dynamics of ranking lists and give some empirical results which will highlight the quite *turbulent dynamics* of cities over time.

In the third part, we will review the various models that have been proposed to describe the temporal evolution of urban populations. In Chapter 5, we will remind the reader of some general results of *stochastic calculus* and stochastic differential equations that are necessary for a good understanding of the mathematical models of urban growth. In Chapter 6, we will present these models, which are *models of preferential attachment*, all related to Gibrat's law of growth. We will present several models which can pertain to city growth, even if they were originally developed through various different phenomena. Finally, we will introduce Gabaix's model. In Chapter 7, we will note that these models concentrate on how cities grow but not why. We will shed light on the particular role played by *interurban migrations* in city growth, as noticed by some authors since the 1970s.

Based on this intuition, in the fourth part we will discuss a different model of urban dynamics starting from first principles that govern the evolution of urban populations. This last part will start with complementary mathematical results whose intuition will be sketched at a more introductory level in Chapter 8. These results will deal with the notion of *rare events*, through the generalized central limit theorem and Lévy stable laws, which will be paramount in the model discussed here. With these tools, starting from first principles and empirical results, in Chapter 9 we will derive the *growth equation of cities*, an equation of a new kind in urban sciences. As we will prove in Chapter 10, this equation will not only represent a quantitative change but a qualitative shift, in the spirit of complex systems. Empirically valid, able to reproduce the turbulent dynamics of cities observed in history, both quantitatively and qualitatively different from what was thought of urban growth before, this equation is what we think should be the new starting point of research in the field of urban population.

Acknowledgments

MB thanks all the colleagues from various horizons—ranging from statistical physics to geography that allowed him to understand different parts of cities: D. Aldous, E. Arcaute, A. Arenas, M. Batty, H. Berestycki, L. Bettencourt, P. Bordin, J.-P. Bouchaud, J. Bouttier, A. Bretagnolle, M. Breuillé, G. Caldarelli, A. Chessa, V. Colizza, Y. Crozet, M. De Domenico, S. Derrible, A. Flammini, S. Fortunato, M. Fosgerau, E. Frias, R. Gallotti, J. Gleeson, M. C. Gonzalez, M. Gribaudi, J. Le Gallo, R. Le Goix, J. Gravier, E. Guitter, D. Helbing, R. Herranz, M. Kivela, P. Krapivsky, R. Lambiotte, V. Latora, F. Le Nechet, M. Lenormand, D. Levinson, J. Lobo, T. Louail, J.-M. Luck, K. Mallick, C. Mascolo, Y. Moreno, I. Mulalic, J.-P. Nadal, V. Nicosia, J. Perret, S. Porta, M. A. Porter, J. Portugali, R. Prieto-Curiel, D. Pumain, F. Radicchi, J. J. Ramasco, J. Randon-Furling, P. S. C. Rao, C. Ratti, S. M. Reia, F. L. Ribeiro, D. Rybski, C. Roth, C. Rozenblat, M. San Miguel, F. Santambroggio, A. Schadschneider, S. Solomon, E. Strano, E. Taillanter, M. Tomko, S. V. Ukkusuri, A. Vespignani, A. Vignes. Last but not least, MB thanks his loving family: Catherine, Esther and Rebecca.

VV thanks the different people who gave some of their time to help him understand complex systems and the problems mentioned in this book. This includes K. Mallick, J. Moran, D. Chavalarias, V. Chomel, A. Pierrot, D. Merigoux, A. Delanoë, C. Ratti, P. Santi, C. Heine, X. Gabaix and J.-P. Bouchaud.

Contents

COUNTING PEOPLE

1   **Urban population**   3
    Defining the city   3
    An historical example: Paris   4
    Functional and morphological definitions   5
    Gridded Population of the World   10
    Discussion   11

2   **Why does population matter?**   13
    Population is a good start   14
    Scaling in cities   16

RANKING CITIES

3   **The distribution of urban populations**   35
    Power-laws   35
    Zipf's law for cities   39
    How to fit a power-law   42
    Revisiting Zipf's law for cities   51

4   **Dynamics of ranking**   54
    Stable versus unstable ranking   54
    Modeling the ranking dynamics   59
    Rank variations of cities   63

MODELS OF URBAN GROWTH

5   **Stochastic calculus**   67
    Brownian motion   67
    Itô and Stratonovich prescriptions   69
    Fokker–Planck equation   73

6   **Stochastic models of growth**   76
    The Yule–Simon model of growth   77
    Gibrat's law for cities   88
    Gabaix's model   92

7   **Models with migration**   94
    A modified Yule–Simon model   94
    A master equation approach   95

Diffusion with noise: The Bouchaud–Mézard model   96

HOW CITIES TRULY GROW

8 The generalized central limit theorem and Lévy stable laws   103
The central limit theorem and its generalization   103
Lévy stable laws   108
The generalized central limit theorem   110

9 From first principles to the growth equation   112
Building a bottom-up equation   113
Gravitational model   117
Minimal model for the interurban migration flows   118

10 About city dynamics   124
Solving a new kind of equation   125
Analysis and scaling of the solution   132
Rank dynamics   134

11 Outlook: Beyond Zipf's law   138
Zipf's law: The end?   138
And space?   139

References   142

Index   150

Counting people

1
Urban population

It is hard to mention urban population without explaining what "urban" means and what cities are. If we all have a basic intuition of what a city is and what it is not, studying cities requires them to be given a more proper definition.

One natural definition is the administrative one: every country can define the boundaries of its different cities and declare that the rules inside are different from the rules outside. In that respect, the word "urban" itself comes from the name of the city that was probably the most prominent in European history: the ancient Rome or the *Urbs*. The boundary of Rome, the *pomerium*, was sacred and gave a legal definition that separated the *Urbs* from the countryside (*ager*). Being inside the *pomerium* had legal and religious effects: the law inside was different from the law outside.

The Roman *pomerium* had the same limitations that contemporary cities can face with regard to their administrative definition. It did not prevent people from settling behind the limits or Rome from sprawling. If the *pomerium* was expanded several times over a long period, there was always a difference between the *pomerium* (the border of the law regime), the walls (that protected the city) and the actual living area.

The same process has occurred in modern history, since at least the Industrial Revolution. The administrative boundaries of cities, usually defined by the territories of the nobility, were overtaken as the suburbs became important. Most of the time, the suburbs became even more important than the original historical city, often leading to a 'rescaling' of the city definition: the medieval London known as the "city of London" is now just 0.2 percent of Greater London.

From a scientific point of view, a sounder and more flexible (and if possible universal) definition of the concept of city is hence necessary to allow for relevant analysis and cross-country comparison. Such a definition should rely on where people live or work rather than intangible limits.

Unfortunately, at this stage, there is no harmonized definition across continents and countries, and in this chapter we will discuss some of the most important definitions that are commonly used.

Defining the city

At a very conceptual level, a city is a spatial concentration of individuals, activities, networks and infrastructures. This concentration allows for interactions, specialization and other benefits. So far, close proximity has been necessary for interaction in particular. Obviously, if new tools allow people to interact independently of distance, this new relation to space could deeply reshape cities. However, cities still exist and are part of a larger system of cities at the country level, but also at a global level (Bretagnolle *et al.*, 2009).

At a more practical level, a city—or a metropolitan area—is intuitively made of an urban core—that is, the city center—and a commuting zone, which is a less urbanized area surrounding the center but functionally dependent on it. From a labour-market perspective, this dependency is usually represented by daily commuting trips between the city and its outskirts (Duranton, 2015), a perspective that is also chosen by national institutes when defining metropolitan areas known as Metropolitan Statistical Areas (MSAs) in the United States (US) and Functional Urban Areas (FUAs) in the European Union (EU) and the Organisation for Economic Co-operation and Development (OECD) (US Office of Management and Budget, 2010; Dijkstra and Poelman, 2012).

Under such a definition cities naturally outgrow their administrative boundaries (Adams *et al.*, 1999), which often does not reflect urban patterns of mobility or social cohesion anymore. However, administrative boundaries remain inescapable, as the usual statistical indicators (population, economic or political indices, for example) are generally collected at the administrative level. This is why metropolitan areas as defined by national institutes usually rely on the substrate of administrative cells. Based on these conditions, metropolitan areas as defined since the 1960s (Adams *et al.*, 1999) have been very useful for comparing cities, at least at a national level.

The usual definitions of metropolitan areas used by national institutes raise many questions. Firstly, they are arbitrary in the sense that they use parametric thresholds (Duranton, 2015) for population or commuting clustering. This results in major differences between national institutes. For example, in the US, the FUAs produced by the OECD and the MSAs (produced by US national agencies) do not overlap, and sometimes display strong differences. This makes cross-country comparison between cities very difficult, if not impossible (Bosker *et al.*, 2021; Dijkstra *et al.*, 2021). Secondly, defining metropolitan areas from commuting trips requires very precise census data on individual mobility, whose quality and accessibility may vary a lot from one country to another (Duranton, 2021).

An historical example: Paris

In order to illustrate the problem of how to define a city, we will consider the example of Paris (Fig. 1.1). The city of Paris has been continuously defined by its walls, which varied over centuries as the city grew. These walls were primarily built to defend the city from its enemies. Successive walls were built over the centuries either by extending existing walls or by demolishing the older ones and replacing them at another location, thereby expanding the city. Parts of these ancient walls have been recovered, including a Gallo-Roman wall, medieval walls, the *Fermiers généraux* wall, built for tax purposes, and the last one, the Thiers wall. The construction of the Thiers wall was proposed by the statesman (and future president) Adolphe Thiers and took place between 1841 and 1846.

These different walls had an important influence on the structure of modern Paris. The *Grands Boulevards* replaced the older Charles V and Louis XII walls, the outer

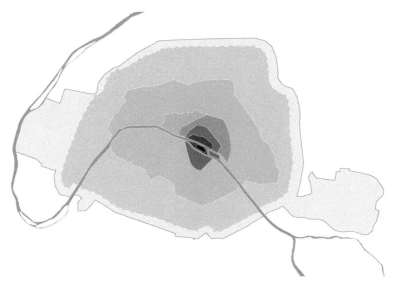

Fig. 1.1: The city limits of Paris from the fourth century to 2015. From dark to light, we show here the Gallo-Roman wall, the first medieval wall, the Philip Augustus wall, the Charles V wall, the Louis XIII wall, the wall of the *Fermiers généraux*, the Thiers wall and finally Paris today. Figure from Wikimedia Commons CC BY-SA 4.0 (user: Paris 16).

boulevards replaced the *Fermiers généraux* wall. The *Boulevard des Maréchaux*, which is a loop encircling the city, was built to replace the Thiers wall, and just beyond it, one can find the ring road called *Boulevard périphérique* (the space between the *Boulevard des Maréchaux* and the *Boulevard périphérique* corresponds to the glacis beyond the Thiers wall). The actual administrative definition of Paris corresponds to this last Thiers wall and encloses what is still today called *Paris intra-muros* (literally "Paris inside the walls").

The city of Paris grew, and the suburbs expanded far beyond the administrative boundaries, as shown in Fig. 1.2. If we focus on the built area, it is then difficult to consider the administrative boundary as still meaningful, apart from the political and social aspects. This example is particularly instructive, as it shows that the administrative origin can have a clear physical origin, but also that it can be completely outdated by the growth of the city and its spatial expansion. This is very clear with this example of Paris, where its administrative division based on the Thiers wall contains about two million residents (2019) in an area of the order of 100 km^2. If we include the suburbs of Paris, the metropolitan area obtained contains approximately 13 million inhabitants on a surface of the order of 19,000 km^2.

Functional and morphological definitions

With time, new proposals and techniques for city delineation have emerged. These fall into two categories: morphological and functional (flow-based) (Duranton, 2021).

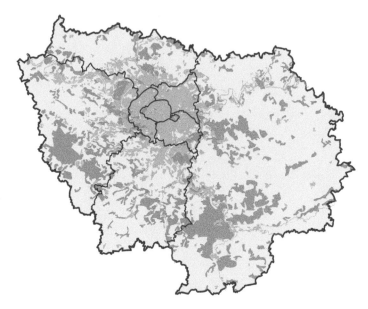

Fig. 1.2: The Paris region now. In this map, the built areas are shown in pink, the forests in green, agricultural zones in yellow and rivers in blue. The thick brown lines show the limit of the region of Ile-de-France. Figure from Thibault Pelloquin (Wikimedia Commons CC BY-SA 3.0).

Morphological approaches define cities as continuous clusters of high population, built-up areas (Arribas-Bel *et al.*, 2021; de Bellefon *et al.*, 2021) or density. They depend on a typical threshold above which we consider the cluster to be a city. The most naive choice in that sense is to define a city as a cluster of more than X inhabitants. The choice of X is, however, arbitrary, since city perception changes a lot through time or from one country to the other. Such definitions also tend to neglect interurban interactions, a core feature of cities.

In contrast, functional definitions rely on flows (Duranton, 2015). All areas that exchange a certain percentage of people everyday with a core center are part of the same city. Mixed approaches combine both categories: FUA and MSA definitions typically define an urban core morphologically and a commuting area functionally (Dijkstra and Poelman, 2012; Moreno-Monroy *et al.*, 2021).

In the following, we will elaborate on the definition of FUAs and MSAs before studying more modern attempts of defining cities.

Functional urban areas: The EU–OECD definition

In order to compare cities and to allow for cross-country analysis, the OECD and the EU jointly proposed a harmonized definition of cities (Dijkstra and Poelman, 2012): the FUA. This definition allows for the identification of 1,320 cities with an urban center of at least 50,000 inhabitants, including 828 cities in Europe and 492 cities in Canada, Mexico, Japan, Korea and the US.

Following this definition, half of the European cities have an urban center with a population in the range 50,000–100,000, and only two of them are global cities (London and Paris). In Europe, these cities host about 40 percent of the EU population, while 30 percent of people are hosted in towns and suburbs that are not included in this city definition. Finally, each city is part of its own commuting zone, and together they form the "Larger Urban Zones" (LUZ) which account for 60 percent of the EU population.

The definition proceeds in four steps. The first step is to identify an urban center based on high-density population grid cells (Fig. 1.3) with a density larger than 1,500 inhabitants per km^2. The second step consists in selecting clustered high-density cells

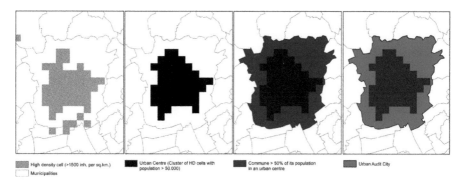

Fig. 1.3: Definition of a city center based on the case of Graz (Austria). We show here the process of going from large density cells to the "urban center". Figure from (Dijkstra and Poelman, 2012).

and keeping clusters with a minimum population of 50,000 as urban centers. The third step is to select all municipalities (local administrative units of level 2) with at least half of their population inside the urban center. Finally, in step four, the city is defined such that there is a link at the political level between selected entities, that at least 50 percent of the city population lives in an urban center and that at least 75 percent of the urban center population lives in a city (Fig. 1.3). We note that in most cases (including the example shown in Fig. 1.3), this last step is not necessary, as the city consists of a single municipality that covers the entire urban center and the vast majority of the city residents live in that urban center.

Once an urban center has been defined, a commuting zone is identified based on commuting patterns and the following requirements: if 15 percent of employed persons live in a city and work in another one, these cities are considered to be one and the same. Municipalities surrounded by a single functional area are included and non-contiguous municipalities are dropped, and the LUZ consists of the city and its commuting zone (see Fig. 1.4).

Metropolitan statistical areas: The US definition

The definition adopted by the US Office of Management and Budget (2010) is in the same spirit as the FUA. The first step is to define a set of "Core Based Statistical

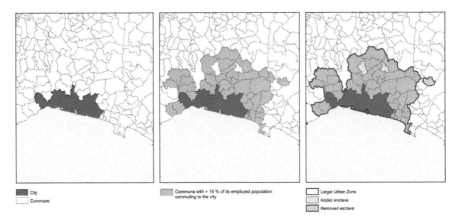

Fig. 1.4: Definition of a commuting zone illustrated using the case of Genova (Italy). Figure from (Dijkstra and Poelman, 2012).

Areas"—the CBSAs (the equivalent of urban centers in the FUA)—which are essentially composed of contiguous counties with relatively high population densities. The counties containing the core urban area constitute the central counties of the CBSA, and at least 50 percent of their population live in urban areas of at least a population of 10,000.

Additional counties are included in the CBSA if they have strong ties to the central counties, measured by commuting and employment. More precisely, if 25 percent of the workers living in a county work in the central counties (or conversely if 25 percent of the employment in the county is held by workers living in central counties), the county is included in the CBSA.

The CBSAs are subdivided into MSAs which are formed around urban areas of at least 50,000 inhabitants and micropolitan statistical areas which are built around urban clusters of populations in the range of 10,000–50,000. Some metropolitan areas may include multiple cities of less than 50,000 people, but combined have over 50,000 people. This minimum of 50,000 for an urban area population to be an MSA might be increased in the future. We show in Fig. 1.5 the various CBSAs of the United States for 2020.

A percolation definition

The previous definitions proposed by the EU–OECD and the US unfortunately rely on a number of parameters that are subjective, and other values would lead to other shapes of the same city. In addition, if population or daily commuting are typical ways of defining cities, more elaborate approaches can be used, especially when trying to define cities where census results are not reliable. Cities can be defined by data sensing, especially satellite images of built-up areas, night lighting (Dingel *et al.*, 2021), roads density (Cao *et al.*, 2020), and so on.

A non-ambiguous definition proposed by scientists relies on the contiguity of built-up areas (Rozenfeld *et al.*, 2011). These authors introduce a percolation-like definition

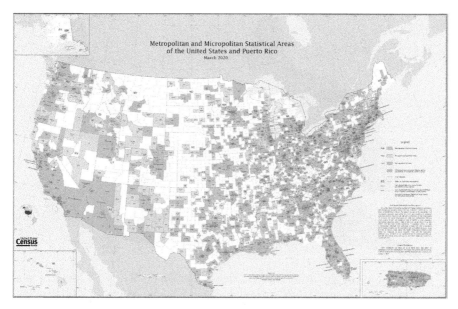

Fig. 1.5: Metropolitan and Micropolitan Statistical Areas (CBSAs) of the US and Puerto Rico (2020). Figure from the US Census Bureau (Public Domain).

and construct cities by clustering areas from high-resolution data. The proposed City Clustering Algorithm (CCA) is defined as follows.

Firstly, a populated area is located and the cluster is grown by adding all populated sites within a distance smaller than some value ℓ and with a population density larger than a threshold ρ^*. The cluster stops growing when no site at a distance less than ℓ and with density larger than ρ^* can be added. In (Rozenfeld *et al.*, 2011), the authors chose ρ^* and took ℓ in the range $[0.2, 4]$ km. It is interesting to note that this procedure leads to clusters that are not too different from the usual MSAs (see Fig. 1.6, for example).

Elaborating on these various ideas of percolation and functional areas, Arcaute *et al.* (2015) proposed a variant of the CCA algorithm to define cities. In a first step, they define urban cores by using population density as the main parameter. They start from a given unit of agglomeration ("wards" in the case of UK). For a given unit, they cluster all adjacent units with density larger than a threshold ρ^* (see Fig. 1.7).

Interestingly, there is a kind of "percolation transition" indicating the emergence of a "giant component" (Stauffer and Aharony, 2018) for a value of $\rho^* \simeq 14$ persons per hectare, above which most cities merge together (Liverpool and Manchester, for example), thus producing a nation-sized cluster which contains the majority of the total population (> 70 percent) and the majority of the total area (> 50 percent). This threshold gives a typical and natural value of density, paving the way to a non-parametric definition of urban centers.

In a second step, to hop from a city-center definition to a metropolitan area one, Arcaute *et al.* (2015) incorporate the notion of functionality through the number of commuters. Once they have determined the urban cores by their density, they consider them as destinations of commuter flows (as long as their population is larger than a

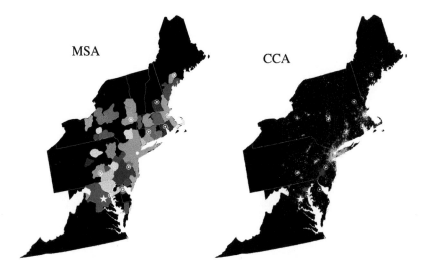

Fig. 1.6: MSA for the counties surrounding New York City (US) and the clusters obtained by the CCA algorithm with $\ell = 5km$. For this value of ℓ, the overlap in terms of population of clusters is maximal. Figures from (Rozenfeld *et al.*, 2011).

threshold N in order to select important commuting hubs only) and they add areas that are the origins of flows. For every given ward, they compute the fraction of individuals who commute to each of the destination clusters and the ward is added to the cluster that receives the largest flow if the flow is above a threshold τ_0. This procedure allows cities to be constructed with only two parameters (N, τ_0). In the remaining part of their paper, they use this definition to test the sensitivity of the usual scaling laws in cities, showing that most deviations from the linear regime are extremely mild, but that exponents can fluctuate considerably when a nonlinear scaling is observed.

Gridded Population of the World

The Gridded Population of the World (GPW) project (NASA, 2022) has the goal of providing a spatially disaggregated population layer that is compatible with other datasets from the social, economic and earth sciences disciplines or from remote sensing. This spatially explicit data is a necessary ingredient for a consistent approach to empirical analysis and policy-making.

This GPW collection, which is regularly updated, is obtained from various population inputs captured at the most detailed spatial resolution available, arising from housing censuses. It is then interpolated for years between censuses. The raster datasets are constructed from national or subnational input administrative units. The gridded count is then obtained by using a proportional allocation gridding algorithm applied to approximately 13.5 million national and subnational administrative units, and assigns population counts to 30 arc-second pixels (which corresponds to approximately 1km at the equator). In Fig. 1.8 we show the global population count consistent with national censuses and population registers for the years 2000 to 2020.

Fig. 1.7: Cities obtained by the clustering algorithm for different density cut-offs. From top left to bottom right: $\rho^* = 40$prs/ha, $\rho^* = 24$prs/ha, $\rho^* = 10$prs/ha and $\rho^* = 2$prs/ha. Figure from (Arcaute *et al.*, 2015).

The GPW can be a very useful population data source when comparing urban population to other urban features. It is not, however, a primary source, since the population data come in the first place from local censuses at the administrative level.

Discussion

As a conclusion, we see that no definition is perfect and that a universal definition of cities would probably make no sense. The definition of a city is intimately linked to the problem considered. In particular, it depends on the level of resolution of the study. Macroscopic studies focusing on migratory flows or international exchanges favour flow

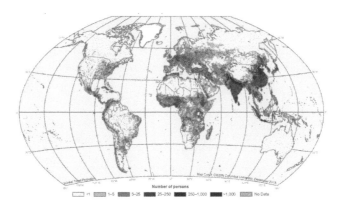

Fig. 1.8: Gridded Population of the World (version 4). Figure copyright: The Trustees of Columbia University in the City of New York (CC BY-SA 4.0).

approaches. Conversely, the understanding of urban sprawl or land use is more related to the notion of built-up areas.

A useful definition must be sufficiently flexible to be able to interpolate between a morphological approach and a functional approach easily, which is possible by varying the value of the thresholds considered (Arcaute *et al.*, 2015). Above all, the most important thing is that everyone uses the same definition in the same context, and, with this in mind, it is probably better to focus on existing definitions, harmonizing them and extending them to new countries, than to look for a new, hypothetically better one. In the remaining part of this book, we will deal mostly with FUAs, which, although not perfect, give a definition of cities that is relevant to different countries.

2
Why does population matter?

Most individuals in the world are now living in cities, and urbanization is expected to keep increasing in the near future (Fig. 2.1). This trend even exists on a much larger time scale, as can be seen in Fig. 2.2. The resulting challenges of this increasing urbanization are complex, difficult to handle and range from the increasing dependence on energy to the emergence of socio-spatial inequalities and serious environmental and sustainability issues. Understanding and modeling the structure and evolution of cities is hence more important than ever, as policy makers are actively looking for new paradigms in urban and transport planning.

Concerning the population of cities, there is a wide variety of sizes, from small towns with 10,000 inhabitants to megacities with populations of more than 10 million.

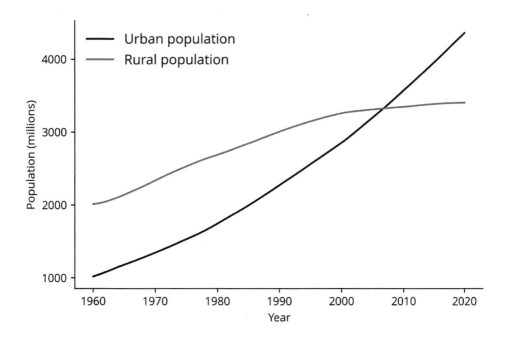

Fig. 2.1: Urban population became dominant at the beginning of the twenty-first century. Yet, on average, the rural population on earth did not decrease. But most population growth has been absorbed by cities.

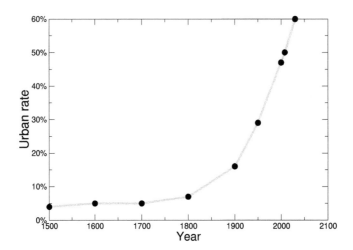

Fig. 2.2: Evolution of the global urban rate (data from the HYDE history database http://themasites.pbl.nl/tridion/en/themasites/hyde).

Large cities and in particular megacities are dominant and represent almost 10 percent of the world's urban population (Fig. 2.3).

Such a variety of possible cities raises many questions: Why are cities so different? How can we compare big and small cities? Are big cities scaled-up versions of smaller ones? Can we say that big cities are more (or less) efficient than small ones? In this chapter, we will see why the study of urban population—the main topic of this book— is at the first order relevant to the understanding of cities. This will naturally lead us to the discussion of *scaling laws*, a very common tool, though not always flawless, in urban studies.

Population is a good start

A natural approach when studying a system is to express its behavior in the form of a transfer function f, which relates the "input" X to the "output" Y

$$Y = f(X), \tag{2.1}$$

where these quantities Y and X are macroparameters that describe the state of the system. The most classical input variable is the size S of the system (to be defined in each case), which allows us to study its scalability: if we double the system size, how does its product behave?

If the size of a city can be understood in several ways—urban area, population, GDP, political importance, etc.—its simplest definition is undoubtedly the population: a city without inhabitants, or an abandoned city, has no meaning even if it occupies space (Louf, 2015). It is therefore very relevant to compare macroscopic urban

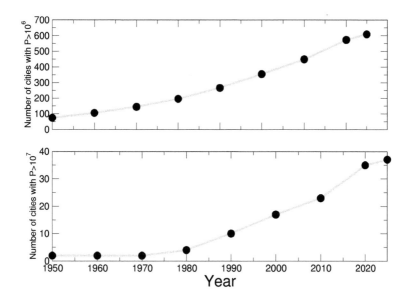

Fig. 2.3: Evolution of the number of cities in the world with a population larger than one million (top) and of megacities with $P > 10^7$ inhabitants (bottom). Data from the UN.

characteristics—economic, social or environmental—with the population. Moreover, when comparing several cities according to a particular characteristic Y, it is common to express the comparison as Y/S, the comparison variable being not Y but Y/S, the product per capita.

This product per capita is constant in the simplest case of linear systems. A system that is twice as large "produces" twice as much

$$f(\lambda X) = \lambda f(X). \tag{2.2}$$

This linear relation can then be used as a baseline that can discriminate between efficient and inefficient cities. A very familiar example, although not always relevant (Arcaute *et al.*, 2015), is the number of patents. Efficient cities produce more patents per inhabitant than the average city, hence having a higher number of patents per capita. Stating that bigger cities produce more patents would result in a relation that is "more" than linear. Studying such relations is the purpose of *scaling laws* that in the urban context were initiated by Pumain, a geographer, and Bettencourt and West, both physicists (Pumain, 2004; Bettencourt *et al.*, 2007).

Scaling in cities

Scaling

Scaling laws and associated scaling exponents are fundamental objects. Used, for example, in biology in order to understand how the metabolic rate varies with body size (Kleiber *et al.*, 1932; Kleiber, 1947), scaling was also widely used in physics to understand polymers (De Gennes, 1979), phase transitions (Goldenfeld, 1992), fluid dynamics and turbulence (Barenblatt, 1996). Scaling also became a central tool for describing the macroscopic properties of complex systems (West *et al.*, 1997*b*; Pumain, 2004; Bettencourt *et al.*, 2007) for two main reasons. First, the existence of a scaling law points to self-similarity: the system reproduces itself as the scale change. Second, scaling laws often behave as power-laws and the corresponding exponents also constitute precious guides for identifying critical factors and mechanisms in complex systems. In particular, when they cannot be deduced from simple dimensional considerations, they point to relevant scales, ingredients, and mechanisms.

Given the simplicity of scaling measures, it is tempting to use this approach in order to get a first grasp of the behavior of complex systems that in general comprise a large number of constituents that interact with each other over various spatial and temporal scales. This is particularly true for urban systems, for which we now have an abundance of data but are still lacking quantitative models for many aspects (Batty, 2013; Barthelemy, 2016; Barthelemy, 2019*a*). For cities, the scaling problem is to understand how extensive quantities vary with the size of the city, usually measured by its population (Pumain, 2004; Bettencourt *et al.*, 2007). Although our theoretical discussion is very general and could in principle be applied to any system, we will use here the language of cities and apply our method to urban data. We thus consider a macroscopic quantity Y that describes a given aspect of cities, which can be socio-economical, about infrastructures, etc., and ask how it varies with the population S of the city according to a given definition. Empirical results for various quantities in cities were compiled for the first time in (Bettencourt *et al.*, 2007) and provided evidence that many quantities follow the scaling relation

$$Y = aS^\beta, \tag{2.3}$$

where a is a prefactor and the exponent β is in general positive. This relation implies that the quantity per capita behaves as $Y/S \sim S^{\beta-1}$, and in the linear case $(\beta = 1)$ the quantity per capita is independent of the size of the city. This is in contrast with all the other cases $(\beta \neq 1)$, where Y/S depends on S, which means that interactions in the city have a nonlinear effect. It is therefore crucial to distinguish the case $\beta = 1$ from $\beta \neq 1$, as it will determine how we model and understand the city. In the seminal paper (Bettencourt *et al.*, 2007), it was shown that we have three different classes of quantities according to the value of β that correspond to different processes. As we just noted, $\beta = 1$ is the linear case, for which the size of city has no impact—think of human-related quantities such as water consumption, for example—while $\beta < 1$ amounts to decreasing returns and $\beta > 1$ to increasing returns or interactions with a positive effect (as expected for creative processes, for example). This study by Bettencourt et al. triggered a very large number of subsequent works, for instance for road properties (Samaniego and Moses, 2008), green space areas (Fuller and Gaston, 2009), urban supply networks (Kühnert *et al.*, 2006), CO_2 emissions in cities (Fragkias *et al.*, 2013; Rybski *et al.*, 2017; Oliveira *et al.*, 2014; Louf and Barthelemy, 2014*b*), interaction activity (Rybski *et al.*, 2009), wealth, innovation and crimes (Bettencourt *et al.*, 2010; Lobo *et al.*, 2013; Alves *et al.*, 2013; Nomaler *et al.*, 2014; Strano and Sood, 2016; Caminha *et al.*, 2017), and so on. These different results motivated the search for a theoretical understanding and modeling that can explain these values (Pumain *et al.*, 2006; Bettencourt, 2013; Bettencourt *et al.*, 2013; Louf and Barthelemy, 2014*a*; Louf and Barthelemy, 2014*b*; Verbavatz and Barthelemy, 2019; Molinero and Thurner, 2019) (see also the review (Ribeiro and Rybski, 2021)).

The idea that a large city is a scaled-up version of a smaller city is encoded in the hypothesis of scaling (Pumain, 2004; Bettencourt *et al.*, 2013). This is a strong assumption that should be carefully scrutinized. In principle, it allows the prediction of a property of one city based on the knowledge of another city and the ratio of their populations. The important underlying assumption, which is intuitively reasonable, is that the population is a good determinant of cities, that is—cities with similar sizes share the same properties. The scaling hypothesis therefore leads to the idea that various aspects of cities can be explored by examining how a quantity Y (which is usually extensive) varies with the population S (Pumain, 2004; Bettencourt *et al.*, 2013). In addition, we note that there are two ways to study this problem. In a 'vertical" approach, one could follow the evolution in time of a city and study how $Y(t)$ obtained for different times varies with $S(t)$. The "horizontal" approach (which is the most common one) consists in compiling Y_i and S_i for various cities i and study the corresponding scaling. It is by no means obvious that these two approaches give the same result, and the study (Depersin and Barthelemy, 2018) showed on the example of congestion-induced traffic delays that the path dependency is so strong that these delays depend on both the population and the whole history of the system, prohibiting the existence of simple scaling forms. It is thus important to bear in mind that scaling for cities is not a trivial fact.

Non-trivial values of β conceal the existence of important mechanisms and could naturally serve as a guide for theoretical models. In this spirit, Bettencourt (2013) suggests that socio-economical quantities (such as wages or patents) increase superlinearly, with an exponent of the form $1 + \delta$ where δ, depends on the fractal dimension of individual paths in the city (for most cases, Bettencourt showed that $\delta = 1/6$). These theoretical predictions agree with various empirical measurements (Bettencourt, 2013), but not all of them (Arcaute *et al.*, 2015; Leitao *et al.*, 2016). In addition, this theoretical approach is of a somewhat phenomenological nature, and it is difficult within its framework to connect microscopic ingredients and mechanisms to the collective emerging behavior. A model that makes it possible to understand and test the effect of various ingredients and mechanisms is still missing (Barthelemy, 2019*b*).

Problems with fitting

The usual (and simplest) way to determine an estimator $\hat{\beta}$ of the scaling exponent β is the ordinary least square regression, which consists essentially in plotting Y versus S in log–log and finding a power-law fit (that is linear in log–log) such that the error (measured as the sum of squared differences) is minimal. This is the classical method used throughout many different fields, which suffers from severe limitations (see Chapter 3). It remains, however, very common and heuristically accepted if the following conditions are satisfied:

(i) the number of decades is sufficient on both axes

(ii) the noise is small

(iii) we are interested in the existence of nonlinearities (which is the case for cities), another constraint is that the exponent should be clearly different from 1.

Unfortunately, however, these conditions are not always met. As an example, we plot the gross domestic product (GDP) for each city in the US (for the year 2010) versus population, and the result is shown in Fig. 2.4. We basically have two decades of

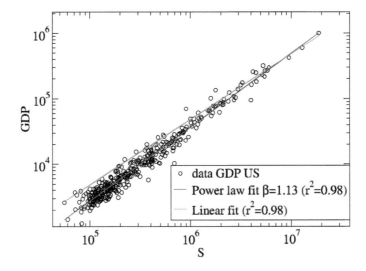

Fig. 2.4: GDP (in millions of current dollars) in 2010 for Metropolitan Statistical Areas (MSAs) versus their population. We show here both the linear and the nonlinear fits. At this point, it is difficult to conclude anything regarding a possible nonlinear behavior. Data from the Bureau of Economic Analysis.

variation on both axes (which, roughly speaking, is the minimum in order to determine a power-law exponent) and a reasonable amount of noise, leading to a power-law fit that gives $\hat{\beta} \simeq 1.13$ (with $R^2 = 0.98$). We see here that conditions (i) and (iii) are not met, and we can only place a relative confidence in the value 1.13. Indeed, a linear fit, which has one parameter less than the nonlinear fit, is also good (Fig. 2.4). More generally, involved statistical methods need to be invoked to know if we can reject the linear assumption or not, and this was the point of the excellent paper by Leitao *et al.*

(2016) (see also Chapter 3 for more proper fitting methods). These authors tested the hypothesis that observations are compatible with a nonlinear behavior, and their conclusion for various quantities is that the estimate of β together with confidence intervals depends a lot on fluctuations in the data and how they are modeled. It is thus difficult to get a clear-cut answer to the fundamental question if β is different from one or not. These fitting problems were also discussed in (Shalizi, 2011) on GDP and income in the US, where it was argued that other scaling forms could be used and that non-trivial scaling exponent values could be an artifact of using extensive quantities instead of intensive ones (per-capita rates).

These problems were reinforced by other studies (Oliveira *et al.*, 2014; Arcaute *et al.*, 2015; Cottineau *et al.*, 2017) that showed the importance of the definition of cities (see also Chapter 1). In (Arcaute *et al.*, 2015), the authors developed a framework for defining cities using commuter numbers and population density thresholds and could show using a UK dataset that many urban indicators scale linearly with population size, independently of the definition of urban boundaries. For quantities that display a nonlinear behavior, the scaling exponent value fluctuates considerably and, more importantly, can be either larger or less than 1 according to the definition used (a problem also observed in the case of CO_2 emissions by transport (Louf and Barthelemy, 2014*b*)). In addition to these empirical problems, we also mention a study on congestion-induced delays in cities which seem to scale with an exponent that varies in time, posing *in fine* the problem of mixing different cities at different stages of their evolution (Depersin and Barthelemy, 2018).

From an Occam's razor perspective, choosing between a linear behavior independent of urban boundaries and a nonlinear scaling exponent whose value fluctuates considerably, leads to the conclusion that many socio-economical indicators are better described by a linear behavior with $\beta = 1$. This is, however, not a scientific proof, and as our capacity for understanding cities relies crucially on this exponent value, the question is still somehow open and begs for a more satisfying answer. As most statistical frameworks and approaches lead to conclusions that critically depend on assumptions, especially in "gray cases" (with large noise, few decades for fitting, an exponent value close to one, etc.), it would be useful to get other evidence of nonlinearities and to somehow circumvent the fitting problem. A value of β that is different from 1 is not only a matter of numerical value, but essentially points to nonlinear effects that are in general relevant at a large scale. In particular, nonlinearities could probably be seen in the dynamics of these systems, and this search could constitute an interesting direction for future (urban) studies. In the following, we will focus on the much more pragmatic question that lies at the core of the idea of (urban) scaling: Knowing some quantity Y_1 for a city of size S_1, what can we say about the corresponding quantity Y_2 for a city of size S_2? In other words, if we accept the idea of scaling, what is the exponent β that we should use in order to compute Y_2 according to $Y_2 = Y_1 (S_2/S_1)^{\beta}$?

The hypothesis of scaling triggered a large amount of research activity that is ongoing, and also raised a number of questions (Bettencourt *et al.*, 2013) that are yet to be answered conclusively (an interesting criticism of scaling in cities was given in (Dyson, 2018)).

Testing the scaling: A statistical approach

The dependence of the exponent β on various parameters such as the city definition led scientists to discuss the likelihood of nonlinear behaviors characterized by $\beta \neq 1$. Rather than finding the exact value of the exponent β, the paramount task is to determine whether β is significantly different from 1 or not. When β is different from 1, one has to investigate the specific processes that can lead to such a value. On the contrary, a linear behavior (with $\beta = 1$) is connected to a trivial behavior (indeed $\beta = 1$ implies that the value per capita Y/S is constant and independent of the city size).

In their 2016 work, Leitao *et al.* used a statistical framework based on a probabilistic formulation of the scaling law which allowed them to perform hypothesis testing and model comparison. They used datasets that included a variety of regions (metropolitan areas of the UK, US, and EU, and others belonging to the OECD). For these databases, they studied various indices of economical activity (income, GDP), innovation (number of patents), transportation (miles traveled, number of train stations), access to culture (number of theaters, etc.) and health conditions (AIDS infections, death by external causes).

Leitao *et al.* (2016) use the likelihood of the data generated by different models. They assume that the quantity y of a city of size x is a random variable with probability density $P(y|x)$. The scaling behavior is then interpreted as

$$\mathbb{E}(y|x) = \alpha x^{\beta}, \tag{2.4}$$

where the expectation value is defined as

$$\mathbb{E}(f(y)|x) = \int f(y) P(y|x) \mathrm{d}y. \tag{2.5}$$

Equation 2.4 does not specify the variance of the distribution or the higher moments of $P(y|x)$. Hence, they choose to restrict the space of $P(y|x)$ to distributions whose variance $\mathbb{V}(y|x) = \mathbb{E}(y^2|x) - \mathbb{E}(y|x)^2$ is of the following form:

$$\mathbb{V}(y|x) = \gamma \mathbb{E}(y|x)^{\delta}. \tag{2.6}$$

This relation between the variance and the mean is also called Taylor's Law (Taylor, 1961), and seems to appear in many complex systems (Eisler *et al.*, 2008). The value of the exponent δ is in general in $[1, 2]$ (in the original study by Taylor, the exponent values are in the range $[0.5, 3]$).

The model is specified by the choice of $P(y|x)$, which has to be consistent with Eqs. 2.5 and 2.6. In (Leitao *et al.*, 2016), the authors use five different models. Two are Gaussian distributions of the form

$$P(y|x) = \frac{1}{\sqrt{2\pi\sigma(x)^2}} e^{-(y-\mu(x))^2/2\sigma(x)^2}, \tag{2.7}$$

where $\mu = \alpha x^{\beta}$ and $\sigma^2(x) = \gamma(\alpha x^{\beta})^{\delta}$. The authors also consider two different models of log-normal distribution. These distributions are prior, and Leitao *et al.* (2016) also consider a fifth model where processes for the assignment of y given x are consistent with Eqs. 2.5 and 2.6.

Once a choice of $P(y|x)$ has been made, we consider each realization of N data points (x_i, y_i) as independent of each other and construct the log-likelihood

$$\ln \mathcal{L} = \ln P(y_1, y_2, \ldots, y_N | x_1, x_2, \ldots, x_N) = \sum_{i=1}^{N} \ln P(y_i | x_i). \tag{2.8}$$

Using 15 different datasets, Leitao *et al.* (2016) test these five different models and perform the following operations:

- They estimate the value of β by maximizing $\mathcal{L}$ over the parameters $(\alpha, \beta, \gamma, \delta)$. They also estimate the error bar around this estimate (using a bootstrapping method (Leitao *et al.*, 2016)).

- For each model, they compute a p-value that tells us if the fluctuations observed in the data are consistent with the model (and whether the residuals are correlated). In other words, they test the hypothesis that the data were generated by the model, and they decide that the model is not rejected if the p-value is larger than 0.05. They also select the model with the lowest Bayesian information criterion (BIC) as the model that best describes the data.

- They quantify the evidence for $\beta \neq 1$ by comparing the maximum likelihood of each model with the same model where $\beta = 1$ is fixed. They construct an indicator Δ that shows if the model with $\beta \neq 1$ provides a better description of the data than with $\beta = 1$. If $\Delta < 0$, the model with $\beta = 1$ is better; if $0 < \Delta < 6$, the evidence for $\beta \neq 1$ is inconclusive; and for $\Delta \geq 6$ the model with $\beta \neq 1$ is better.

- They quantify the evidence for a non-trivial scaling in Taylor's law ($\delta \neq 1$ or $\delta \neq 2$).

They obtain the following results: in 3 of the 15 cases, at least one model is compatible with the data. In the best model, the value found for β allows a comparison between $\beta \neq 1$ and $\beta = 1$. They obtain that the UK income and UK train stations display a behavior that is linear with population, while the GDP (for OECD countries) is most likely nonlinear.

In the remaining 12 cases, none of the five models is compatible with the data. However, for 8 of the 12 remaining datasets, different models, although not compatible with the data, give similar answers. This is the case for UK patents and OECD patents displaying a linear behavior; it is also the case for USA-GDP, EU museums, Brazil GDP with $\beta > 1$, and USA roads, EU libraries and Brazil AIDS cases with $\beta < 1$. The analysis is inconclusive in the remaining four cases as to whether $\beta \neq 1$ or $\beta = 1$, since the value of β varies too much (from above to below 1) depending on the model.

The fact that almost all models are rejected by the data shows that finding improved models is crucial. The estimation of β and the development of generative models go together, and it appears essential to consider the predicted fluctuations not only in the validation of the model but also in the estimation of β.

A "practical" test of scaling

The main idea.   Instead of fitting the data considering all cities at the same time—with all the limitations discussed above—we consider two cities, 1 and 2, with populations S_1 and S_2. Assuming a scaling form Eq. 2.3 to be correct (with the standard assumption of a constant prefactor; see (Bettencourt, 2021) for a generalization of the scaling form) and knowing S_1, S_2, and Y_1, we obtain Y_2 as

$$Y_2 = Y_1 \left(\frac{S_2}{S_1} \right)^{\beta}. \tag{2.9}$$

The scaling assumption and the fitted value of β allow us to predict the behavior of a scaled-up version of a given city $\hat{Y}_2(\beta)$. Conversely, we could also ask what would be the "local" exponent that allows us to correctly predict Y_2. Obviously, we have from Eq. 2.9

$$\beta_{\mathrm{loc}} = \frac{\log(Y_2/Y_1)}{\log(S_2/S_1)}, \tag{2.10}$$

which has the simple geometric interpretation of being the slope of the straight line joining the points (S_1, Y_1) and (S_2, Y_2) in a log–log representation. In the absence of noise and if all data points are aligned, there is a unique value β_{loc} for all pairs of cities which corresponds to the value $\hat{\beta}$ obtained by the direct fit. Let r be the population ratio S_2/S_1, where we consider that S_2 is the largest population so that we always have $r \geq 1$. In the general case, studying the value β_{loc} tells us how different cities are related to each other giving a representation of scaling across different values of the size ratio, akin to some sort of "tomography" scan of scaling. Plotting β_{loc} versus r is what we will call the "tomography plot," as it allows us to explore scaling for various cross-sections of the size ratio.

If we assume that $Y_2 = Y_1(S_2/S_1)^{\beta}(1+\eta)$ where η is noise, we obtain for $S_2/S_1 > 1$ the general expression

$$\beta_{\mathrm{loc}} = \beta + \frac{\log(1+\eta)}{\log(S_2/S_1)}. \tag{2.11}$$

This expression shows that when the noise is not too large, the effective exponent converges for large S_2/S_1 to the theoretical one and to its estimate via fitting: $\beta_{\mathrm{loc}} \simeq \hat{\beta} \simeq \beta$. This expression also shows that a plot of β_{loc} versus $\log S_2/S_1$ for all pairs of cities should display a hyperbolic envelope and that $\beta_{\mathrm{loc}} \to \beta$ for large size ratio values. For similar populations $S_2 = S_1(1+\varepsilon)$ (with $\varepsilon \ll 1$), we obtain at lowest order in ε

$$\beta_{\mathrm{loc}} \simeq \beta + \frac{\log(1+\eta)}{\varepsilon}. \tag{2.12}$$

We see here that for small ε, we can observe arbitrary large values of β_{loc} for non-zero fluctuations η. For similar cities, noise is therefore relevant, and their comparison cannot help us much in determining the scaling exponent.

Alternatively, if the noise is non-multiplicative, we can construct an expression of the form $Y_2 = Y_1(S_2/S_1)^\beta + \eta$. The noise η cannot be too large, otherwise the scaling assumption is not correct and Y_2/Y_1 does not depend on the ratio S_2/S_1 only. However, if we accept this form, a simple calculation shows that

$$\beta_{\text{loc}} = \beta + \frac{1}{\log r} \log(1 + \frac{\eta}{r^\beta Y_1}), \tag{2.13}$$

which also shows that for large r there is a convergence toward β for a large class of noise η. In particular, for r large enough, we have

$$\beta_{\text{loc}} \simeq \beta + \frac{1}{r^\beta \log r} \frac{\eta}{Y_1}, \tag{2.14}$$

which shows that even in this case, if the scaling assumption is correct, there should be a convergence of β_{loc} toward β.

This local exponent allows us to define and identify a "benchmark city" that can serve as a reference value for computing quantities for other cities. More precisely, for a city i, one can compute the corresponding local exponents for all other cities j versus the ratio $r_{ij} = P_j/P_i$ as

$$\beta_{\text{loc}}(i,j) = \frac{\log(Y_j/Y_i)}{\log r_{ij}}. \tag{2.15}$$

The average and the variance of the local exponent when varying j are

$$\langle \beta_{\text{loc}}(i) \rangle = \frac{1}{N-1} \sum_j \beta_{\text{loc}}(i,j) \tag{2.16}$$

$$\sigma^2(i) = \langle \beta_{\text{loc}}^2(i) \rangle - \langle \beta_{\text{loc}}(i) \rangle^2, \tag{2.17}$$

where the brackets denote $\langle O \rangle = \sum_j O(j)/(N-1)$ (N is the number of cities). We then define the benchmark city such that the variance $\sigma^2(i)$ is the smallest possible, and we label it i_{min}. For this city, the fluctuations of the local exponent are the smallest possible around its average $\beta_{\text{eff}} \equiv \langle \beta_{\text{loc}}(i_{\text{min}}) \rangle$. This city can be seen as a benchmark in the sense that we can use it for computing "reliable" properties of other cities through the formula

$$Y(j) = Y(i_{\text{min}}) \left(\frac{S_j}{S_{i_{\text{min}}}} \right)^{\beta_{\text{eff}}}. \tag{2.18}$$

This justifies the denomination "effective exponent," as it can be used for practical predictions. Other choices for an effective exponent are of course possible, but in the spirit of practical applications, we are interested in picking a single value of β for computing the quantity Y for all cities. In this respect, minimizing the variance of β_{loc} is a simple sensible answer to this question, although probably not the only one.

We note here that this discussion is different from the one about Scale-Adjusted Metropolitan Indicators (SAMIs) defined in (Bettencourt and West, 2010; Lobo *et al.*, 2013) as being the variation of a given city with respect to the fit given by $\hat{\beta}$:

$$\xi_i = \log \frac{Y_i}{Y_0 S_i^{\hat{\beta}}}. \tag{2.19}$$

We will, however, consider a similar quantity. Knowing β_{eff}, we compute the fraction $f(\varepsilon_1, \varepsilon_2)$ of cities for which

$$\varepsilon_1 Y_{\text{data}} < Y_{\text{predicted}} < \varepsilon_2 Y_{\text{data}}, \tag{2.20}$$

where Y_{data} is the actual value for a given city of population S and

$$Y_{\text{predicted}} = Y(i_{min}) \left(\frac{S}{S_{i_{min}}} \right)^{\beta_{\text{eff}}}. \tag{2.21}$$

In particular, we will focus on the case $\varepsilon_1 = 1/\varepsilon_2$ for different values of ε_2. Additional information can be provided by plotting the function $f(\varepsilon) \equiv f(1/\varepsilon, \varepsilon)$ for $\varepsilon > 1$, and we will show it in a few cases (we will also give the value of $f(2)$ for $\varepsilon_2 = 2$, as it gives a good idea of the accuracy of the prediction computed with β_{eff}).

Practical cases. We now apply these tools to the different datasets discussed in (Leitao *et al.*, 2016) and the other datasets mentioned above. These datasets concern different areas of the world (Europe, the US, OECD, Brazil) and various socio-economical quantities, and were analyzed with standard statistical methods. Therefore they represent an interesting benchmark for testing other methods. We note that for most of these datasets, cities have to be understood as urban areas, except for Brazilian data, where administrative boundaries were used. We first discuss cases for which there is little or no ambiguity about the scaling behavior, and see how it is confirmed with the tools proposed here. We then focus on less clear cases for which we find results not completely consistent with the classical analysis. Finally, we will elaborate on the datasets where the statistical analysis in (Leitao *et al.*, 2016) was "inconclusive," meaning that the result was depending on the assumption taken for the disorder. Our main goal here will be to show how these tools can shed new light on these problematic or inconclusive cases.

We consider here the two datasets for the US that were studied in (Leitao *et al.*, 2016). The first one is about the GDP of cities and the second about the number of miles of roads (in each city). The nonlinear fits for these two quantities are shown in Fig. 2.5. In the first case, the GDP displays a clear superlinear behavior with $\hat{\beta} = 1.11$, while for infrastructure the expected sublinear behavior is observed with $\hat{\beta} = 0.85$. We now inspect in more detail these cases with the help of the local exponent β_{loc} (see Fig. 2.6). We observe on these plots that the "naive" nonlinear fitting is confirmed: for most pairs of cities, the local exponent is different from one and is equal to $\hat{\beta}$ (within error bars). If we now compute the effective exponents, we obtain for the GDP $\beta_{\text{eff}} = 1.13 \pm 0.07$ and for the number of miles $\beta_{\text{eff}} = 0.80 \pm 0.1$. We note that in both cases the benchmark city is New York City (NYC), the largest urban area

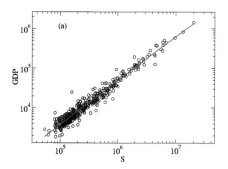
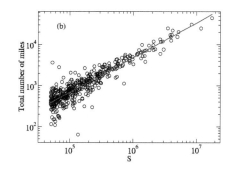

Fig. 2.5: (a) GDP (in millions of current US dollars) for cities in the US for the year 2013 (Leitao *et al.*, 2016). The red line is a power-law fit with $\hat{\beta} = 1.11$ and $R^2 = 0.98$. (b) Number of miles for US cities for the year 2013 versus population. The power-law fit gives $\beta = 0.85$ ($R^2 = 0.91$).

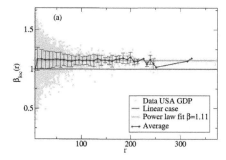
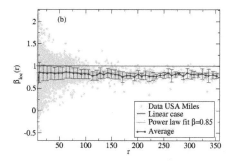

Fig. 2.6: Local exponent versus population ratio r for (a) the GDP for cities in the US for the year 2013 and (b) the total number of miles in the US for the year 2013.

in the US.[1] We see here that all evidences are pointing to the same conclusion of a nonlinear behavior. Even if $\beta = 1.11$ is only slightly different from 1, the tomography plot (Fig. 2.6a) clearly shows the reality of this nonlinear exponent and is confirmed by the value of the effective exponent $\beta_{\text{eff}} = 1.13$. Using these effective exponents for the GDP and the number of miles, the fraction $f(2)$ is 98 percent and 91 percent for the GDP and the miles, respectively. In other words, using NYC as the benchmark city and the effective exponents, we get excellent predictions for all the other cities.

Finally, in order to address the problem discussed in (Shalizi, 2011), we redo the analysis in the US case for the GDP per capita. In this case the nonlinear fit is indeed less good, with an exponent 0.11 ($R^2 = 0.42$) but which corresponds well to the value

---

[1] At this point, we note that there is no clear rule for identifying *a priori* the benchmark city.

$\hat{\beta} - 1$. The tomography plot constructed for this quantity is shown in Fig. 2.7 and seems to be free of any ambiguity: for most values of r, the local exponent is equal on average to 0.11 and converges quickly toward this value when r increases, and is strictly positive (within error bars). In addition, the effective exponent is in this case equal to $\beta_{\mathrm{eff}} = 0.13 \pm 0.07$ and leads to an impressive value of the fraction of cities $f(2) = 98\%$, whose value is correctly predicted. All these elements suggest that there is indeed a nonlinear behavior for the GDP in US cities, even if we work on the GDP per capita that should exclude effects due to the extensivity of this quantity, as claimed in (Shalizi, 2011). There could be large fluctuations, and the local exponent

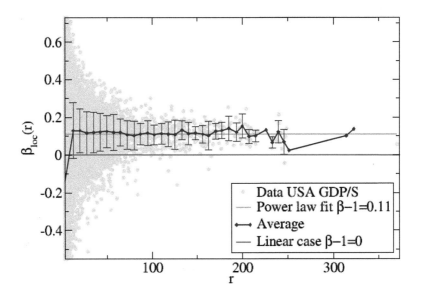

Fig. 2.7: Tomography plot for the GDP per capita for US cities for the year 2013.

analysis reinforces the naive fit in some cases, while in others it casts some doubts on the nonlinear behavior (Barthelemy, 2019b).

We consider now datasets for which the analysis in (Leitao *et al.*, 2016) suggested conclusive results about the linear or nonlinear behavior (although most of these datasets were not compatible with any of the considered models). These datasets are the UK train stations, AIDS cases in Brazil and the number of patents in cities belonging to OECD countries.

The authors of (Leitao *et al.*, 2016) studied the number of train stations in UK cities and found a clear linear behavior, $\hat{\beta} = 1.0$. However, if we plot this number versus the population, we obtain the result shown in Fig. 2.8a. We first observe that there is a

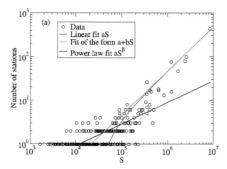
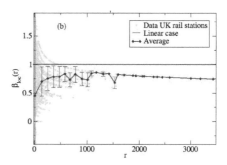

Fig. 2.8: (a) Number of rail stations vs. population in UK cities (Leitao *et al.*, 2016). We show here the linear fit aS with $a \simeq 4.6710^{-5}$ ($R^2 = 0.98$), the fit $a + bS$ where $a \simeq -1.42$ and $b \simeq 4.7210^{-5}$ ($R^2 = 0.98$), and the power-law fit aS^β where $a \simeq 0.01$ and $\beta \simeq 0.50$ ($R^2 = 0.76$). (b) Tomography plot for the number of rail stations.

lot of noise and the quality of any fit will likely be very poor. Also, we note that there is a large number of cities with one station exactly, which potentially will affect any fitting method. Given all these problems, the linear fit is not too bad, in agreement with the result of the analysis of (Leitao *et al.*, 2016). However, the plot of the local exponent versus r shown in Fig. 2.8b signals the existence of important problems. Indeed, this plot seems to indicate a sublinear behavior, far from the linear prediction, but also with very large fluctuations.[2] This inconsistency suggests the existence of a problem in the dataset. The presence of large fluctuations could be a reason for the discrepancy observed between the linear behavior and $\beta_{loc}(r)$, but it could also signal another scaling form. In particular, the data are not inconsistent with a fit of the form $a + bS$, where $a < 0$ (see Fig. 2.8a), implying a threshold effect: for $S < S_c \simeq 30,084$ we have no stations, while for $S \gg S_c$ we observe a linear behavior. In the power-law scaling assumption, we can compute the effective exponent and find $\beta_{\text{eff}} = 0.12 \pm 0.17$ with $f(2) = 85\%$, but given the large level of noise and the high likelihood of another scaling form, we do not assign high confidence to this result.

The situation for the number of AIDS cases in Brazil for the year 2010 is similar to the previous case. The plot of this number versus population is shown in Fig. 2.9a. The power-law fit gives an exponent $\hat{\beta} = 0.74$ consistent with the sublinear conclusion of (Leitao *et al.*, 2016), but given the large fluctuations, a fit of the form $a + bS$ is also consistent with the data. This last fit predicts a threshold effect with $S_c = 10,090$ and a linear behavior for $S \gg S_c$, similar to the previous case of UK rail stations. The tomography plot (Fig. 2.9b) shows that the scaling behavior is not clear with a range around $r \sim 10^3$, for which the local exponent is close to 1, but for other values of r we observe a sublinear exponent. The effective exponent is $\beta_{\text{eff}} = 1.03$ with a fraction

---

[2]The different hyperbolas appear because of cities with the same number of stations, such as one or two stations. We added noise to the data in order to destroy this effect and observe that the tomography plot is robust.

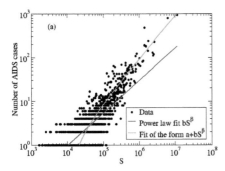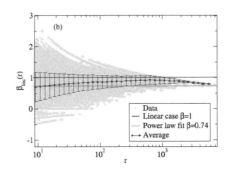

Fig. 2.9: (a) Number of AIDS cases versus city population in Brazil for the year 2010. We show the power-law fit with exponent $\hat{\beta} = 0.74$ ($R^2 = 0.81$) and the fit of the form $a + bS^{\hat{\beta}}$ with $\hat{\beta} = 0.99$, $a = -1.009$, and $b = 0.00010$ ($R^2 = 0.93$). (b) Corresponding tomography plot.

$f(2) = 67\%$. It thus seems here that the sublinear conclusion of (Leitao *et al.*, 2016) could actually be challenged by a threshold function and/or a linear behavior.

For some of the datasets studied in (Leitao *et al.*, 2016), standard tools could not lead to a clear conclusion about whether the scaling is linear or not. There are two main reasons for this. The first, (i), that β can be larger or smaller than 1 depending on the assumption used for describing the fluctuations. The second reason, (ii), is that for some cases the best model for fluctuations only marginally improves the statistics compared with the linear fit. The relevant datasets of (Leitao *et al.*, 2016) are the following: for Europe, the cinema capacity and usage (ii) and the number of theaters (i), and for Brazil the number of deaths caused by external causes (ii). These cases therefore represent interesting playgrounds for testing other methods. We will use the tools developed above and will show that our method can bring some new conclusions or a new perspective, such as the existence of a threshold.

We start with the cinema capacity (total number of seats) in European cities. The naive fit gives a linear behavior with $\hat{\beta} = 0.99$ ($R^2 = 0.71$). The tomography plot confirms this: there is a convergence of β_{loc} to 1 (see Fig. 2.10a). The calculation of the effective exponent gives $\beta_{\mathrm{eff}} = 0.98$, and for this value the fraction of cities with a prediction in $[0.5, 2.0]$ is $f(2) = 74$ percent. All these results point in favor of a linear behavior. Even if the statistical evidence found in (Leitao *et al.*, 2016) for this behavior seemed not to be large enough, we have here an objective 74 percent of cities whose cinema capacities are correctly predicted using an exponent equal to 0.98.

In the case of cinema usage computed as the attendance in cinemas in the year 2011, the power-law fit gives the exponent $\hat{\beta} = 1.46$ ($R^2 = 0.64$), indicating a strongly non-linear behavior. The tomography plot is seen in Fig. 2.11a and shows that for most pairs of cities the local exponent is larger than 1, except for very large ratios $r \gtrsim 60$. This suggests that there is a tendency toward a nonlinear behavior in agreement with the value of the effective exponent, which is $\beta_{\mathrm{eff}} = 1.17$. For this value, however,

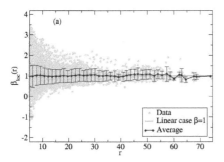
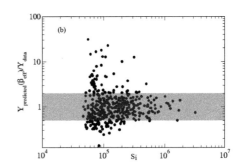

Fig. 2.10: Cinema capacity in European cities. (a) Tomography plot and (b) ratio $Y_{predicted}/Y_{data}$ versus the population. The gray area represents the fraction of cities with ratio in the range of $[0.5, 2.0]$, which is about 74 percent here.

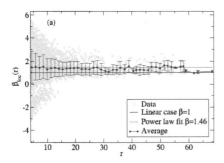
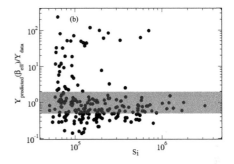

Fig. 2.11: Cinema usage in European cities for the year 2011. (a) Tomography plot. (b) Ratio $Y_{predicted}/Y_{data}$ versus the population. The gray area represents the fraction of cities with ratio in the range $[0.5, 2.0]$, which here is about 50%.

the ratio $Y_{predicted}/Y_{data}$ (shown in Fig. 2.11b) indicates large fluctuations with only about 50% of cities with a ratio $Y_{predicted}/Y_{data}$ in the range $[0.5, 2.0]$. The other 50 percent display a ratio in the range $[0.1, 0.5]$ or a much larger one, up to 10^2. In this respect, with such large fluctuations it is indeed a bit hard to conclude, although the superlinear behavior with $\beta_{\text{eff}} = 1.17$ accounts for half of the cities.

We now focus on the number of theaters in European cities for the year 2011. This case was classified as inconclusive in (Leitao *et al.*, 2016), as the exponent value for β could be either larger or smaller than 1 depending on the assumptions about the fluctuations. Despite large fluctuations, we can try a power-law fit and the corresponding exponent is $\hat{\beta} = 0.91$ ($R^2 = 0.74$) (Fig. 2.12a). The tomography plot (Fig. 2.12b) confirms that this is indeed a difficult case: for $r \lesssim 40$, the local exponent is around 1, while for larger values we observe local exponents smaller than 1 and even smaller than $\hat{\beta}$. There is, therefore, no clear convergence toward the fitting value, and this might explain why this case, despite showing a relatively clear sublinearity, was considered inconclusive in (Leitao *et al.*, 2016). This forces us to reconsider the validity of the power-law fit, knowing that we have essentially one decade of variations, which is far from being enough for a good fit. We note here that a fit of the form $a + bS$ (or obviously a more complex one of the form $a + bS^{\beta}$), with $a = -0.51$ and $b = 2.510^{-5}$ ($R^2 = 0.68$), is also consistent with data. This last fit implies a threshold value of $S_c \simeq 20,400$, above which the number of theaters is non-zero. We note that a threshold effect is here somehow expected; indeed, only for large enough cities do we observe the appearance of theaters. If we try, however, to compute the effective exponent, we obtain $\beta_{\text{eff}} = 0.95$ and the corresponding fraction is $f(2) = 71\%$, suggesting a slightly sublinear behavior here. This effective exponent together with the tomography plot, therefore, suggests a slightly sublinear behavior, but we cannot exclude the possibility of a threshold effect (since these are not mutually exclusive properties).

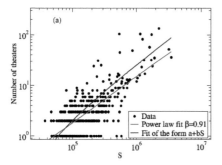 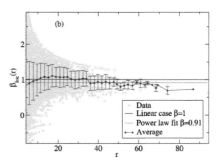

Fig. 2.12: (a) Number of theaters in European cities versus their population. The red line is the power-law fit with exponent $\hat{\beta} = 0.91$ ($R^2 = 0.74$). The green line is the fit of the form $a + bS$ with $a = -0.51$, and $b = 2.510^{-5}$ ($R^2 = 0.68$). This fit implies a threshold value $S_c \simeq 20,400$. (b) Tomography plot: local exponent versus population ratio r for the number of theaters in European cities.

Finally, we consider the dataset of death by external causes in Brazil. This database is provided by Brazil's Health Ministry for the year 2010. In this case, too, the authors of (Leitao *et al.*, 2016) found that there was not enough statistical evidence to be able to come to a conclusion. We show this number versus the population in Fig. 2.13a for various fits. The power-law fit is not too bad, and predicts a linear behavior $\hat{\beta} = 1.03$. The forms $a + bS$ and $a + bS^{\beta}$ do not produce consistent results about the existence of a threshold effect: for the linear fit there is no threshold, while for the second fit (of the form $a + bS^{\beta'}$) there would be a small threshold value $S_c = 6,500$ (and $\beta' = 0.90$). It is therefore hard to conclude at this stage, but the tomography plot shown in Fig. 2.13b is rather clear and points to a linear behavior: the local exponent converges quickly toward 1 and its average is equal to 1 for all values of r (within error bars). The

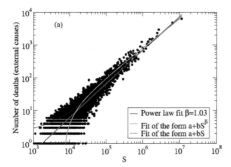 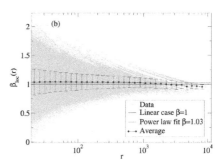

Fig. 2.13: Death by external causes in Brazil for the year 2010. (a) Number of deaths versus the size of population. We show here various fits including the power-law fit aS^β with $\beta = 1.03$ ($R^2 = 0.91$), and fits of the form $a + bS$ (with $a = 2.31$, $b = 0.00068$, $R^2 = 0.97$) or $a + bS^\beta$ (with $a = -9.40$, $b = 0.0035$, $\beta = 0.90$, $R^2 = 0.98$). (b) Tomography plot showing a convergence toward an exponent equal to 1.

effective exponent computed for this case is $\beta_{\text{eff}} = 0.99$, and the fraction of cities whose number of deaths is correctly predicted with this value is about 82 percent. Despite the difficulties with fitting the original data, we have here an interesting case where the local exponent analysis is clear and all evidence points to a linear scaling.

Discussion. We have discussed simple tools for analyzing data that could help with understanding their scaling behavior. Although these tools do not replace the standard statistical analysis, they enable a more practical view of the system's behavior: if we had to use the scaling form for making predictions, what would be the most reliable exponent? One advantage of this approach is that the answer to this question does not depend on some assumptions, such as the nature of the noise. In cases where noise is small and the number of available decades is large, this analysis simply confirms standard tools such as fitting methods. It is in more complex cases where it is difficult to decide which model describes the data the best, that this type of method could be of some help. The analysis of the local exponent gives a precise picture of how different systems of different sizes are related to each other. In some cases it allows us to reach conclusions about the nonlinear or linear behavior with more confidence, but in other cases it also signals the failure of a simple scaling. This failure could happen due to a threshold effect, for example, but more generally, we could expect the system to be described by a more complex function, with more than one exponent, for example.

We can summarize this analysis by proposing the following set of necessary conditions in order to trust the fitting value $\hat{\beta}$:

- We need the convergence of β_{loc} toward $\hat{\beta}$.
- The value of the effective exponent β_{eff} should be consistent with $\hat{\beta}$. In general, the value β_{eff} should be preferred over $\hat{\beta}$, in particular if the value $f(2)$ is large.
- The value of $f(2)$ should be at least 50 percent, which could, of course, be debated, but at least we should observe a rapid increase of $f(\varepsilon, 1/\varepsilon)$ with decreasing ε.

If these conditions are not satisfied, we can safely reject the value $\hat{\beta}$ obtained by the power-law fit. In this case, it suggests, for example, either that fluctuations are too large or that the simple power-law scaling form is not valid.

In the next chapter, while investigating the proper ways to fit the population distribution of cities, we will keep elaborating on power-laws. In particular, we will discuss the best way to a fit a power-law in a general context. Some of the results we will derive will give complementary insight to the scaling discussion of the present chapter.

Ranking cities

3
The distribution of urban populations

It is remarkable that cities are not all of the same size. This seemingly trivial fact disqualifies the idea of an "optimal size" of cities, a typical size scale that would reflect optimal transport costs or living conditions for people. On the contrary, there is no rule that leads cities to a universal equilibrium—typically between economical increasing and decreasing returns to scale—and universal size, as would result from urban economics models. More importantly, the most recognized and characterized quantitative urban fact is the idea of cities having a regular and universal *inequality of sizes*, long approximated by Zipf's law (Zipf, 1949). This paradox is what Krugman (1996) called the "mystery of urban hierarchy." Understanding urban population and urban growth therefore implies understanding how and why cities can have different populations, first by measuring these differences and then by explaining them. From a dynamical point of view, another question is also to see if an equilibrium distribution for cities makes sense, i.e. if it means something to study cities at equilibrium in some static way.

This chapter aims to describe how the distribution of urban population has been considered so far. After some theoretical reminders on power-laws, we will see how urban populations could appear to be described by Zipf's law, which we will formally introduce. Then, after further reminders of power-law fitting methods, we will return to Zipf's law assumption for cities by empirically proving that Zipf's law is, in general, not true for cities.

Power-laws

Power-laws in social phenomena

When thinking of statistical distributions, the normal law strikes us as one of the most frequent marks of universality in mathematics. This is, in a sense, expected, as the normal law is the attraction law of most of the sums of independent identical random variables.[1] It could be expected, then, that any summation of (reasonably bounded) complex events should, in one way or another, lead to the emergence of a normal law. This is why many empirical quantities cluster around a typical value with possible but rare deviations. An empirical quantity y following a normal law of mean μ and deviation σ is of the order of this mean:

---

[1]But not all sums: see Chapter 8.

$$y \simeq \mu \pm \sigma. \tag{3.1}$$

In the nineteenth century, the normal law served as the main conceptual framework for statistical thought (Desrosières, 2016). The normal law appeared to be the natural law of equality: It defined a standard ideal, like Quetelet's average man, around which variations or errors were necessarily small.

However, all distributions do not fit the pattern of a typical value and can exhibit significant deviations. For such cases, there is no typical value with which the empirical quantity can be approximated, and the average value can be far from the most likely value. The interest in these kinds of "broad" distributions grew at the end of the nineteenth century when statistical thinking evolved and considered the tails of distributions. In a partially eugenicist vision stemming from Darwinism, statisticians switched from considering the average size, strength or wealth to the tallest, strongest and richest. No longer was it the average individual that counted but the best one, capable of surviving the law of evolution, in society as in nature. This is how Francis Galton, Charles Darwin's cousin, introduced the concept of regression: by comparing the height of children with that of their parents, he found that there was on average a linear relationship between the height of children and that of their parents, but that the linear coefficient of this line was less than 1 and therefore tended to bring the height of the tallest children back to the mean: it was therefore a *regression*, taken in the sense of degeneration.

For statisticians at the beginning of the twentieth century, tails mattered more than bells. Gradually, even in social sciences, the standard statistical distribution shifted from the egalitarian normal law to unequal ones that emphasized the difference between individuals, like power-laws. Vilfredo Pareto was the first to observe a power-law in a social phenomenon: the distribution of income in the population. On this subject, Benoît Mandelbrot commented:

The data did not remotely fit a bell curve, as one would expect if wealth were distributed randomly. "It is a social law," [Pareto] wrote: something "in the nature of man."

> Benoît Mandelbrot – *The Misbehavior of Markets: A Fractal View of Financial Turbulence*
> (2006) (Mandelbrot and Hudson, 2007)

Pareto's interpretation led him to assert that a superior *élite* always and necessarily emerges from a group of people. It is interesting to see that even today the appearance, sometimes real, often alleged, of power-laws in economics can serve as a justification for Darwinist interpretations or *laissez-faire*. Yet, it remains an inescapable framework in modern applied statistics, whose definition needs to be reminded.

Definition

Power-laws have attracted attention over the years for their mathematical and physical properties. Formally, a power-law is a relationship between two quantities of the form (for x larger than some positive value)

$$y = bx^{-\alpha} \tag{3.2}$$

where b is a constant of proportionality and a scaling exponent $\alpha > 0$ (and a similar relation when x is negative and large).

The main characteristic of power-laws is their *invariance of scale*. Whatever the scale of the variable x, the system is qualitatively identical. If we multiply x by a constant c

$$y(cx) = b(cx)^{-\alpha} = c^{-\alpha}y(x) \propto y(x). \tag{3.3}$$

In particular, the ratio $y(x)/y(z)$ depends only on the ratio x/z:

$$\frac{y(x)}{y(z)} = \Phi\left(\frac{x}{z}\right). \tag{3.4}$$

The Pareto law is a particular type of power-law, in the form of a probability distribution. Let X be a random variable, and X follows a Pareto distribution if

$$\mathbb{P}(X > x) = \left(\frac{x_{\min}}{x}\right)^{\alpha} \text{ for } x \geq x_{\min}, \tag{3.5}$$

where α is the Pareto exponent and $x_{\min}$ a minimal value. The distribution of X is

$$f(x) = \alpha\frac{x_{\min}^{\alpha}}{x^{\alpha+1}} \text{ for } x \geq x_{\min} \tag{3.6}$$

and its moments of order n are finite for $\alpha > n$. In particular, a Pareto law with exponent $\alpha < 2$ has no variance and one with exponent $\alpha < 1$ has no mean.

Pareto's law defines power-law distributions in their most restrictive form. Here, we choose a looser definition of power-law distributions: a power-law distribution is a distribution which, from a certain point, behaves as a power-law for large values:

$$p(x) \underset{x \gg 1}{\sim} x^{-1-\alpha}. \tag{3.7}$$

These distributions exhibit several interesting properties. Their main property is to be connected to a generalized version of the central limit theorem (see Chapter 5). If the X_i are independent random variables identically distributed according to a power-law distribution of exponent α, we have the following convergence in law

$$\sum_{i=1}^{n} X_i \overset{\mathcal{L}}{\longrightarrow} \begin{cases} (bn)^{1/\alpha}\zeta_{\alpha} & \text{if } \alpha < 1 \\ n\mu + (bn)^{1/\alpha}\zeta_{\alpha} & \text{if } 1 < \alpha \leq 2 \\ n\mu + (bn)^{1/2}\eta & \text{if } 2 < \alpha, \end{cases} \tag{3.8}$$

where μ is the mean (when it exists), ζ_{α} is a Lévy stable law of parameter α (see Chapter 8) and η a Gaussian variable.

From a physical point of view, the difference between a power-law distribution and another decreasing function (such as an exponential or Gaussian, for example) is not only quantitative but also qualitative. Indeed, for the exponential (or similarly the Gaussian), there is a characteristic scale above which the probability is essentially negligible. In contrast, for power-laws there is no such scale and in fact events of several order of magnitude larger than the average can be observed in reality. The existence of such rare events is a typical signature of power-law distributions.

Zipf's law

Zipf's law is a sort of discrete version of Pareto's law, identified empirically at about the same time by Jean-Baptiste Estoup (1916) and George Kingsley Zipf (1949) in the linguistic domain (see Fig. 3.1) and Felix Auerbach (1913) for the distribution of city populations.

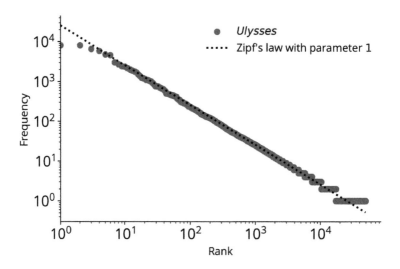

Fig. 3.1: Frequency of appearance of words in the *Ulysses* by James Joyce (1920). The frequency of appearance of words in the novel follows a Zipf's law, as identified by Zipf himself.

In the linguistic domain, Zipf's law states that there is a relationship between the frequency f_n of appearance of a word in a text and its rank n, the word appearing most often being of rank 1. This relationship is of the form

$$f_n \sim \frac{1}{n^s}, \tag{3.9}$$

where the exponent s is close to 1. Depending on sources, Zipf's law is defined by the value $s = 1$, by a value close to 1 or by any value. In this book, we define Zipf's law in a broad sense, where s can take any value, which allows us to define the *Zipf's law of parameter s*.

This relation allows us to link Zipf's law to the discrete Pareto law:

$$f_n \sim \frac{1}{n^s} \iff \mathbb{P}\left(S > f_n\right) \sim n \tag{3.10}$$

$$\iff \mathbb{P}\left(S > \frac{1}{n^s}\right) \sim n \tag{3.11}$$

$$\iff \mathbb{P}\left(S > x\right) \sim \left(\frac{1}{x}\right)^{\frac{1}{s}}, \tag{3.12}$$

These relations show that Zipf's law of parameter s is related to the Pareto law of parameter $1/s$. In other words, if Zipf's law is of parameter s, the corresponding probability distribution behaves as a power-law with exponent $1 + 1/s$.

Different mathematical models have been proposed to explain the emergence of Zipf's law in social phenomena (see Chapter 6). In the following, we are only interested in Zipf's law for cities. For a more in-depth description of the appearance of power-laws in social or economic phenomena, we can recommend reading (Moran, 2020).

Zipf's law for cities

Auerbach–Zipf's law

Zipf's law in the urban domain originated before Zipf, in an article by the German physicist Felix Auerbach published in 1913 (Auerbach, 1913). In ranking the largest German cities in 1913 by population, Auerbach observed that the product of the rank of the cities by their population was more or less constant and equal to what he defined as the *Absolute Konzentration (AK)*, the value of which was close to 50 (see Fig. 3.2). Thus, for all n, we have

$$nS(n) \simeq \text{AK}, \tag{3.13}$$

where n is the rank of the population city $S(n)$, which corresponds well to Zipf's law with exponent 1.

Auerbach also demonstrated that his law was valid in time—i.e. for several periods of German history—and in space—i.e. for different countries. Hence, he was the first to produce an empirical observation of Zipf's law and the first to make a comparative analysis of its appearance.

The regularity in time and space of Zipf's law, which was observed many times after Auerbach (Singer, 1936; Rosen and Resnick, 1980; Krugman, 1996; Blank and Solomon, 2000; Ioannides and Overman, 2003; Corominas-Murtra and Solé, 2010), is probably the most striking quantitative fact—not to say the only one—of the science of cities, to the point of being a landmark of many urban models (Duranton, 2021): a model is good if it reproduces Zipf's law.

This result characterizes the hierarchical organization of cities and, in particular, quantifies the statistical occurrence of large cities. Zipf's law implies that in any country, the most populous city is generally twice as large as the next largest, and so on. This is a signature of the very high heterogeneity of city size and shows that cities are not governed by optimal considerations that would lead to a single size but, on the contrary, that city sizes are widely distributed and follow a kind of hierarchy (Pumain and Moriconi-Ebrard, 1997). The empirical value of the Pareto exponent thus tells us about the degree of hierarchy of a city system: a large value of the Pareto exponent (and thus a small value of the Zipf exponent) corresponds to a more equal distribution of the population among the cities, while for small values of the exponent, the city system is very heterogeneous and dominated by a few metropolises.

Properties of systems of cities

Consider a country of N cities with city populations S_i distributed according to Zipf's law with parameter s. The population distribution follows a Pareto law with exponent

Nr.	Ort	E.-Z.	A.K.	Nr.	Ort	E.-Z.	A.K.
1	Berlin	3579	36	48	Mülhausen	109	58
2	Hamburg	1168	23	49	Saarbrücken	105	51
3	Leipzig	622	19	50	Hamborn	108	52
4	München	610	24	51	Lübeck	99	51
5	Köln	593	30	52	Beuthen	98	51
6	Dresden	575	35	53	Bielefeld	97	51
7	Breslau	522	37	54	Münster	90	49
8	Essen	475	38	55	Oberhausen	90	50
9	Frankfurt a. M.	449	40	56	Bonn	88	49
10	Nürnberg	414	41	57	Metz	87	50
11	Hannover	383	42	58	Darmstadt	87	51
12	Düsseldorf	358	43	59	Bremerhaven	86	51
13	Elberfeld-Barmen	354	46	60	Görlitz	86	52
14	Stuttgart	315	44	61	Würzburg	84	51
15	Chemnitz	306	46	62	Freiburg i. Br.	83	52
16	Mannheim-Ludwigsh.	282	45	63	Bromberg	80	50
17	Magdeburg	280	48	64	Recklinghausen	79	51
18	Bremen	267	48	65	Wilhelmshaven	78	51
19	Dortmund	263	50	66	Offenbach	76	51
20	Stettin	258	52	67	Waldenburg i. Schl.	75	51
21	Königsberg i. Pr.	246	51	68	Remscheid	72	49
22	Duisburg	229	51	69	Bottrop	71	49
23	Kiel	209	48	70	Pforzheim	69	48
24	Bochum	193	46	71	Frankfurt a. O.	68	48
25	Straßburg	191	48	72	Ulm	68	49
26	Halle	188	48	73	Gera	68	50
27	Gelsenkirchen	177	48	74	Witten	67	50
28	Danzig	170	48	75	Harburg	67	51
29	Kassel	165	48	76	Gleiwitz	67	52
30	Posen	157	47	77	Liegnitz	67	52
31	Aachen	156	48	78	Osnabrück	66	52
32	Krefeld	155	50	79	Rostock	65	51
33	Augsburg	151	50	80	Koblenz	65	52
34	München-Gladbach	146	50	81	Potsdam	65	53
35	Braunschweig	144	50	82	Buer	62	51
36	Mainz	137	49	83	Heidelberg	61	51
37	Zabrze	136	50	84	Flensburg	61	51
38	Karlsruhe	134	51	85	Mörs	60	51
39	Solingen	129	50	86	Elbing	59	51
40	Kattowitz	128	51	87	Herne	57	49
41	Erfurt	124	51	88	Dessau	57	50
42	Plauen	121	51	89	Kaiserslautern	55	49
43	Königshütte	120	52	90	Brandenburg	54	49
44	Wiesbaden	119	52	91	Trier	54	49
45	Zwickau	118	53	92	Regensburg	53	49
46	Mülheim a. R.	112	52	93	Hildesheim	50	47
47	Hagen	110	52	94	Thorn	50	47

Fig. 3.2: First notice made by Auerbach in 1913 on german cities of the fact that rank × population is roughly constant. Table from (Auerbach, 1913).

$\mu = 1/s$ so that

$$P(S) \sim \frac{1}{S^{1+\mu}}. \tag{3.14}$$

The total urban population in the country is

$$U(N) = \sum_{i=1}^{N} S_i, \tag{3.15}$$

which is a sum of independent identically distributed random variables. When $\mu > 1$, the central limit theorem yields

$$U(N) \sim N, \tag{3.16}$$

whereas when $\mu < 1$, from the generalized central limit theorem, the sum is dominated by its largest terms and

$$U(N) \sim N^{1/\mu}, \tag{3.17}$$

since in this case the population of the largest city is

$$S_{\max} \sim N^{1/\mu}. \tag{3.18}$$

More generally, the rank $r(x)$ of a city of size x is typically represented by

$$r(x) = N\mathbb{P}(S > x) = N \int_x^\infty P(x)\mathrm{d}x, \tag{3.19}$$

where

$$P(x) = \frac{1}{N} \frac{S_{\max}^\mu}{x^{1+\mu}} \tag{3.20}$$

to ensure good normalization.

The population S_2 of the second largest city is such that

$$2 \sim N \int_{S_2}^\infty \frac{1}{N} \frac{S_{\max}^\mu}{x^{1+\mu}} \mathrm{d}x \tag{3.21}$$

and

$$2 \sim \frac{S_{\max}^\mu}{S_2^\mu}. \tag{3.22}$$

The ratio between the two largest city populations is thus

$$\frac{S_{\max}}{S_2} = 2^{1/\mu} \tag{3.23}$$

and is called the *primacy index*. The primacy index is close to 2 when Zipf's law exponent is close to 1. The smaller the μ, the more hierarchical the system of cities is, implying that the largest city is more important in the total demographics of the country. We note that in the case of a Gaussian population distribution, the ratio $S_{\max}/S_2$ would be close to 1. The hierarchical distribution of cities is inseparable from a notion of broad distribution.

Finally, we note that the total number of cities with population larger than a value S is given by

$$N_>(S) = N \int_S^\infty P(x)\mathrm{d}x \simeq \frac{N}{S^\mu}. \tag{3.24}$$

which can also be used to extract the value of μ.

How to fit a power-law

One of the most commonly recurring questions of power-laws is how to recognize them. On the one hand, power-laws are easy to spot: as explained in Chapter 2, they correspond to a linear trend in log–log scale. Properly fitting a power-law and finding a good estimate of the Pareto exponent (or scaling parameter) α is, however, a more complex question, and we can rarely be sure that we have identified a power-law (for a similar discussion in the context of complex networks, see (Broido and Clauset, 2019)).

When studying a set of data that could plausibly yield a power-law distribution, there are hence three questions to address:

1. Does the data roughly look like it is power-law distributed? The answer to this question is yes if the data seem to collapse on a straight line in log–log scale, which is easy to verify.
2. What is a plausible value for the Pareto exponent α? Multiple tests can be made to find the most likely value of α, assuming *ex ante* that the distribution is power-law distributed. We will present here four methods: the ordinary least-square regression, the maximum likelihood estimator, the Hill's estimator and the Kolmogorov–Smirnov test. Indeed, a straight line in log–log scale is a necessary, yet not sufficient, condition for a power-law.
3. Are we sure that what we see is a power-law? This is the most difficult question. No dataset will follow a perfect power-law, and the aim is to understand if modeling the dataset by a power-law is relevant in the context of study. Indeed, and this is the case for any statistical problem, it is easier to perform parametric tests (that could give the most plausible α value) than non-parametric tests (that would give the distribution family). Several clues can indicate whether a power-law is irrelevant or sound or if another family of distribution is more likely. Here, we will give some in the context of city population.

The ordinary least-square regression

The easiest way to fit a power-law, and the most often used, dates back to Pareto's work on wealth distribution. Constructing the histogram of the empirical quantity x, one can see in log–log scale if $\log p(x)$ versus $\log x$ collapses on a straight line (see Fig. 3.3), meaning that

$$\log p(x) = \alpha \log x + \beta, \tag{3.25}$$

where β is a constant. The most natural way to estimate the value of α is, then, to perform an ordinary least-square regression (OLS) regression.

The OLS regression is a linear least-square method for estimating unknown parameters. The idea is to minimize the sum of the squares of the differences between the observed variable and those predicted by the linear function. More precisely, if we observe the set $\{x_i, y_i\}$ for $i = 1, \ldots, N$, we want to find the parameters (α, β) of the function $y = f(x) = \alpha x + \beta$ such that

$$\phi = \sum_{i=1}^{N} (f(x_i) - y_i)^2 \tag{3.26}$$

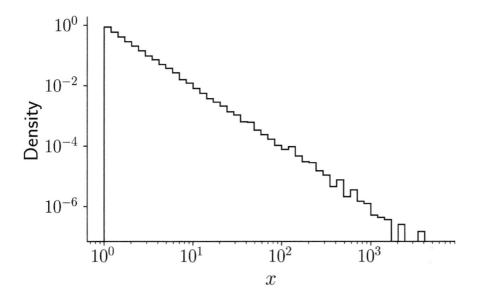

Fig. 3.3: Example of a distribution displaying a power-law behavior, represented by a straight line in log–log scale. The slope gives the exponent.

is minimum. We then have two parameters and the OLS conditions read

$$\frac{\partial \phi}{\partial \alpha} = \frac{\partial \phi}{\partial \beta} = 0, \tag{3.27}$$

which can be easily computed :

$$\langle y \rangle - \alpha \langle x \rangle - \beta = 0 \tag{3.28}$$

$$\langle xy \rangle - \alpha \langle x^2 \rangle - \beta \langle x \rangle = 0, \tag{3.29}$$

where the brackets denote the average $\langle o \rangle = 1/N \sum_i o_i$. We can solve this system of equations for α and β and get the regression coefficients

$$\hat{\alpha} = \frac{\langle yx \rangle - \langle x \rangle \langle y \rangle}{\langle x^2 \rangle - \langle x \rangle^2} \tag{3.30}$$

$$\hat{\beta} = \frac{\langle x^2 \rangle \langle y \rangle - \langle x \rangle \langle y^2 \rangle}{\langle x^2 \rangle - \langle x \rangle^2}. \tag{3.31}$$

How well the OLS regression fits can be determined by estimating how much of the fluctuation in the sample can be reduced with the regression. The R^2 coefficient in $[0, 1]$ is then defined as the ratio of the "explained" variance to the "total" variance of the variable y:

$$R^2 = \frac{\langle (f(x) - \langle y \rangle)^2 \rangle}{\langle y^2 \rangle - \langle y \rangle^2}.$$ (3.32)

Roughly speaking, the larger the value of R^2 and the better the fit.

Maximum likelihood estimator

Another way of looking at OLS from a more probabilistic point of view is to consider that it is equivalent to the maximum likelihood estimator (MLE) of the underlying Gaussian model of the linear regression.

Indeed, in this perspective, we assume that the variables x and y are linked through the relation

$$y = \alpha x + \beta + \epsilon,$$ (3.33)

where ϵ is a Gaussian random variable and $\epsilon \sim \mathcal{N}(0, \sigma)$.

One then wants to estimate which α and β are the most likely to generate the observed sample $\{x_i, y_i\}$. By assumption, $y - \alpha x - \beta$ follows a Gaussian law of zero mean and variance σ. The probability of observing $\{x_i, y_i\}$ knowing α and β is then

$$\mathbb{P}(x_i, y_i | \alpha, \beta) = \prod_{i=1}^{N} \exp\left(-\frac{(y_i - \alpha x_i - \beta)^2}{2\sigma^2} \right).$$ (3.34)

From Bayes' theorem,

$$\mathbb{P}(\alpha, \beta | x_i, y_i) \propto \mathbb{P}(x_i, y_i | \alpha, \beta) \mathbb{P}(\alpha, \beta)$$ (3.35)

where $\mathbb{P}(\alpha, \beta)$ is the prior probability of (α, β). The prior probability translates the prior knowledge we have about the variables α and β (one can, for example, expect that they fall within a typical range of values). Without further knowledge, this prior probability is uniform. In this case,

$$\mathbb{P}(\alpha, \beta | x_i, y_i) \propto \mathbb{P}(x_i, y_i | \alpha, \beta),$$ (3.36)

and we define the log-likelihood as

$$l(\alpha, \beta) = \sum_i \log \mathbb{P}(x_i, y_i | \alpha, \beta).$$ (3.37)

The values of α and β that maximize the log-likelihood are the ones that are the most *likely* in the sense that they maximize the probability $\mathbb{P}(\alpha, \beta; x_i, y_i)$ of α and β knowing the sample values (x_i, y_i).

From Eq. 3.34, we have

$$l(\alpha, \beta) \propto \sum_{i=1}^{N} -(y_i - \alpha x_i - \beta)^2,$$ (3.38)

which is exactly Eq. 3.26 from the OLS equation. The values

$$\hat{\alpha} = \frac{\langle yx \rangle - \langle x \rangle \langle y \rangle}{\langle x^2 \rangle - \langle x \rangle^2} \tag{3.39}$$

$$\hat{\beta} = \frac{\langle x^2 \rangle \langle y \rangle - \langle x \rangle \langle y^2 \rangle}{\langle x^2 \rangle - \langle x \rangle^2} \tag{3.40}$$

are thus the ones that maximize the likelihood of the observed values (x_i, y_i) assuming that Eq. 3.33 holds.

From the MLE, one can also derive the variance of the estimators $\hat{\alpha}$ and $\hat{\beta}$. Indeed, expanding the likelihood function (and dropping β for simplicity) near $\hat{\alpha}$ gives

$$l(\alpha) \simeq l(\hat{\alpha}) + \partial_\alpha l(\hat{\alpha})(\alpha - \hat{\alpha}) + \frac{1}{2} \partial_\alpha^2 l(\hat{\alpha})(\alpha - \hat{\alpha})^2. \tag{3.41}$$

By definition of $\hat{\alpha}$, $\partial_\alpha l(\hat{\alpha}) = 0$. We then define Fisher information as

$$I(\alpha) = -\partial_\alpha^2 l(\hat{\alpha}), \tag{3.42}$$

so that

$$l(\alpha) \simeq l(\hat{\alpha}) - \frac{1}{2}(\alpha - \hat{\alpha})^2 I(\hat{\alpha}). \tag{3.43}$$

The Fisher information measures the amount of information that the observation of a random variable carries about an unknown parameter of a distribution that models this random variable.

By definition of the likelihood, this amounts to

$$\mathbb{P}(\alpha; x_i, y_i) \propto \exp\left(-\frac{(\alpha - \hat{\alpha})^2}{2} I(\hat{\alpha})\right), \tag{3.44}$$

which means that at the first order α follows a normal law of mean $\hat{\alpha}$ and variance $1/I(\hat{\alpha})$ (see Fig. 3.4). Hence, a typical confidence bound for α is:

$$\alpha \in \left[\hat{\alpha} - \frac{1}{\sqrt{I(\hat{\alpha})}}, \hat{\alpha} + \frac{1}{\sqrt{I(\hat{\alpha})}}\right]. \tag{3.45}$$

In the case of the linear regression, from Eq. 3.38 we obtain

$$I(\hat{\alpha}) = \sum_{i=1}^{N} -\partial_\alpha^2 \frac{-(y_i - \alpha x_i - \beta)^2}{2\sigma^2}$$

$$= \frac{1}{\sigma^2} \sum_{i=1}^{N} x_i^2 \tag{3.46}$$

so that we have

$$l(\alpha) \simeq l(\hat{\alpha}) - \frac{N}{2\sigma^2}(\alpha - \hat{\alpha})^2 \langle x^2 \rangle \tag{3.47}$$

and

$$\mathbb{P}(\alpha; x_i, y_i) \propto \exp\left(-\frac{(\alpha - \hat{\alpha})^2}{2\sigma^2} \langle x^2 \rangle N\right). \tag{3.48}$$

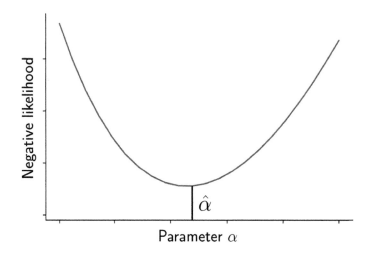

Fig. 3.4: The estimator $\hat{\alpha}$ is the one that maximizes the likelihood. Fisher information $I(\hat{\alpha})$ is the curvature of the likelihood function and a typical standard deviation for the estimation of α is then the radius of curvature $1/I(\hat{\alpha})$.

The typical estimation of α in a linear regression model is then

$$\alpha \simeq \hat{\alpha} \pm \frac{\sigma}{\sqrt{\langle x^2 \rangle}} \frac{1}{\sqrt{N}}. \tag{3.49}$$

From the previous derivation, we saw that the OLS is equivalent to assuming an underlying Gaussian random variable in the linear model. It is, usually, not a big issue. In the case of a power-law distribution of the form $p(x) = Ax^{-\alpha}$, applying an underlying linear model to $\log p(x)$ would, however, amount to saying that the probability $p(x)$ has log-normal fluctuations, which would violate the central-limit theorem (Clauset *et al.*, 2009). An OLS estimation of a power-law exponent is hence theoretically not right, and the previous formulas are inapplicable to the power-law fit. This is the main limitation to the use of OLS in fitting power-laws. It has other empirical limitations like the risk of making distributions that are far from a power-law look like ones with a *spurious exponent*, as we will see later through examples.

Maximum likelihood estimator for power-laws

What we have discussed previously for estimating a power-law exponent with the minimum likelihood is not, however, useless. We will simply use it in a more straightforward way.

Instead of assuming that the log of the considered empirical distribution $p(x)$ follows a linear model (which is theoretically wrong), we will assume that $p(x)$ follows a power-law. On that condition, one can determine the value of Zipf's exponent. Of course, this would not prove that the assumption of the power-law is correct—as stated before, it is almost impossible to prove that an empirical distribution follows a power-law. This, however, gives a valid estimation of the exponent and confidence bound. By investigating the stability and validity of the fit, one can find reasonable pro and contra arguments of the power-law model.

We assume that a random variable X has a continuous distribution of the form

$$p(x) = Cx^{-\alpha}. \tag{3.50}$$

normalized over the range $[x_{min}, \infty)$ which gives the constant C. We then find

$$p(x) = \frac{\alpha - 1}{x_{min}} \left(\frac{x}{x_{min}} \right)^{-\alpha}. \tag{3.51}$$

We assume now that we have N observations $\{x_i\}$ of the random variable X, and we would like to know the value of α that could generate this sample. The probability that the data is drawn from the model with value α is given by

$$p(\{x_i\}; \alpha) = \prod_{i=1}^{n} \frac{\alpha - 1}{x_{min}} \left(\frac{x_i}{x_{min}} \right)^{-\alpha}. \tag{3.52}$$

Following the reasoning of Eq. 3.35 to Eq. 3.37, we find that the log-likelihood function of the power-law model is $l(\alpha) = p(\{x_i\}; \alpha)$, which is given here by

$$l(\alpha) = N \ln(\alpha - 1) - N \log x_{min} - \alpha \sum_{i=1}^{N} \ln \frac{x_i}{x_{min}}. \tag{3.53}$$

Writing $\partial l / \partial \alpha = 0$ then leads to the maximum likelihood estimator:

$$\hat{\alpha} = 1 + \frac{N}{\sum_{i=1}^{n} \log (x_i/x_{min})}. \tag{3.54}$$

The MLE estimator of a continuous power-law of exponent α converges almost surely to α for $N \to \infty$ (Clauset *et al.*, 2009). The Fisher information (see Eq. 3.42) in this case is

$$I(\alpha) = \frac{N}{(\alpha - 1)^2} \tag{3.55}$$

so that the estimator $\hat{\alpha}$ is asymptotically Gaussian with variance $(\alpha - 1)^2/N$. So we can roughly write for large N

$$\hat{\alpha} \simeq \alpha + \xi \frac{|\alpha - 1|}{\sqrt{N}}, \tag{3.56}$$

where ξ is a noise of order $\mathcal{O}(1)$.

MLE for the Pareto exponent is the most natural and robust way of finding the good value of the exponent, providing that we assume that the empirical distribution follows a power-law.

The Hill estimator and Hill plots

Another way of assessing that a distribution follows a power-law, which is a little bit more elaborate than a visual confirmation that the data points collapse on a straight line in a log–log scale, is Hill's method of estimation and Hill's plots. Hill's method is a pseudo-MLE for power-laws (Hill, 1975). We assume now that we have N observations $\{x_i\}$ of the random variable X. We define the k-th order statistic of X as the k-th smallest value of $\{x_i\}$. We then define Hill's estimator for some k_n as

$$\hat{\alpha}_{k_n} = 1 + \frac{1}{\frac{1}{k_n}\sum_{i=1}^{k_n} \log\left(X_{(i)}/X_{(k_n)}\right)}, \tag{3.57}$$

where $X_{(1)} \geq X_{(2)} \geq \cdots \geq X_{(k_n)}$ are the $k_n + 1$ largest order statistics of $\{x_i\}$. When $k_n = N$, one retrieves the MLE of α.

The idea behind Hill's estimator is to perform an MLE on subsets of the total sample starting from the highest values. If the underlying distribution converges to some asymptotic power-law regime, Hill's estimator should converge quickly to some sort of plateau close to the exponent value of the power-law regime. If not all the distribution is a power-law, Hill's estimator should diverge when entering the domain of large-order statistics. Hence, Hill's estimator, which is related to Hill's plot (see Fig. 3.5), is a good way to visually identify a power-law regime in an empirical distribution. If the Hill's plot reaches a plateau over a large domain of order statistics, this indicates the existence of a power-law regime in the underlying distribution.

Kolmogorov–Smirnov estimates

A last possible way of assessing the validity of power-law fit is to use the standard Kolmogorov–Smirnov test (Massey Jr, 1951). The Kolmogorov–Smirnov test is a non-parametric test of the similarity between two probability distributions. More precisely, let two probability distributions have cumulative distribution functions $F(x)$ and $G(x)$; then the Kolmogorov–Smirnov statistics is defined as

$$D = \sup_x |F(x) - G(x)| \tag{3.58}$$

and is the largest "gap" between the two cumulative distribution functions (see Fig. 3.6). Consider an empirical cumulative distribution function for n independent and identically distributed ordered observations of the random variable X and G_α the cumulative distribution function of the Pareto law of exponent α. The empirical distribution function F_n for these n independent observations is given by

$$F_n(x) = \frac{\text{number of elements} \leq x}{n}. \tag{3.59}$$

The Kolmogorov–Smirnov distance between the empirical distribution and the power-law distribution of parameter α is then given by

$$D_\alpha = \sup_x |F_n(x) - G_\alpha(x)|. \tag{3.60}$$

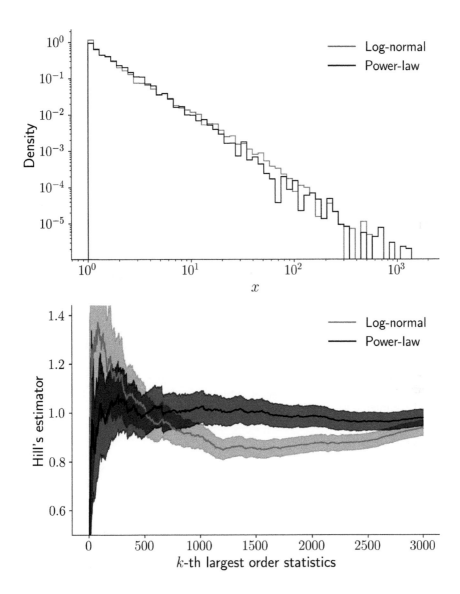

Fig. 3.5: Power-law and log-normal distributions can look very similar in a log–log scale and can be hard to distinguish. However, the Hill's plots of these two distributions are very different: the Hill's plot of the power-law rapidly plateaus to its true exponent value (here $\alpha = 1$), whereas the Hill's plot of the log-normal distribution does not.

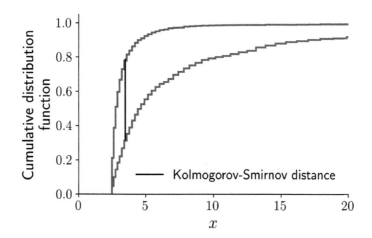

Fig. 3.6: The Kolmogorov–Smirnov distance is the maximal distance between two cumulative distributions.

The Kolmorogov–Smirnov estimate of the parameter α for the empirical distribution is then

$$\alpha = \arg\min_{\alpha} D_{\alpha}. \tag{3.61}$$

It is the exponent value that minimizes the distance between the two cumulative distribution functions.

Kolmogorov–Smirnov estimates of power-law exponents are not sufficient to account for the existence of a power-law or best parameter estimation. However, they provide good complementary evidence in power-law identification.

Spurious exponents

So far, we have seen that if power-law-like distributions are easy to spot—corresponding to a straight-line behavior in a log–log scale—and if there are several techniques to estimate the most probable value of the power-law exponent, it is very difficult to confirm that empirical distributions *truly* have a power-law behavior. On the contrary, many distributions can look like power-laws at first glance (see Fig. 3.5) while being different and having very different properties. Gan *et al.* (2006) show, for a whole series of random simulations following reference distributions (normal, lognormal, Pareto and Gamma), that a simple linear regression of Zipf's law apparently gives extremely satisfactory results (with R^2 very close to 1), even when the underlying distribution has nothing to do with Zipf's law. This last result shows that in general how well Zipf's law fits when applied to an empirical distribution can be a confounding artifact. Often resulting from statistical misinterpretations, Zipf's law does not necessarily require economic theory: it is often *spurious*.

Returning to cities, we will now see that the same theoretical reasons for mistaking certain distributions with power-laws has disproportionately led to the misinterpretation of Zipf's law in the city context.

Revisiting Zipf's law for cities

Meta-analysis: A large diversity of exponent values

Zipf's law, although it still appears as a cornerstone of urban studies—every course in urban economics begins with Zipf's law—has been called into question over the last 20 years by the increase in the number of available urban demographic data (Gan *et al.*, 2006). This questioning has taken place in two stages; firstly, by the measurements of Zipf exponents that were increasingly distant from the central value 1 (Soo, 2005; Cottineau, 2017), and, secondly, by general questioning of the content and validity of Zipf's law (Arshad *et al.*, 2018).

Cottineau carried out a synthetic meta-analysis of Zipf's exponent measurements in the literature for different city systems at different times (Cottineau, 2017). We reproduce the results of this study in Fig. 3.7. Although the study shows that the exponent measured in the literature is most often close to 1, it can vary considerably from one country to another or within the same country between two different periods, without any apparent reason. This study therefore corroborates several previous works which, locally, had already questioned the relevance of Zipf's law, through the empirical variations observed around the exponent 1.

An unsteady result

Since the main claim and quality of Zipf's law is that it is universal, its invalidity, even for a small number of systems, brings into question its global meaning. The estimation method and the definition of the system considered are decisive in the determination of the exponent value. We show, for example, on Fig. 3.8 how the value of the exponent, for the same country, can vary according to the number of cities included in the distribution (by taking the n largest cities). This result shows that the alleged power-law for the city size distribution shifts with the number of cities considered, a quite convincing argument against the existence of this power-law. To confirm this, we also show the corresponding Hill's plots. We see that, except for the remarkable case of France which seems to be robust in exhibiting a power-law with an exponent close to 1, the distribution of urban populations in the considered countries does not obey Zipf's law. France is an exception in following Zipf's law. This shows that if city population can be distributed according to Zipf's law, it is in general not the case: Zipf's law does not exist as a universal result.

How can we explain, then, the apparent ubiquity of Zipf's law? As we have mentioned before, the answer comes from the law itself: it is in fact extremely easy to *fit* Zipf's law on a probability distribution, biased by the choice of cities in the literature: the obstinacy to look for Zipf's law made it easy to prove but in general, there is no reason to think that Zipf's law exists for cities (Benguigui and Blumenfeld-Lieberthal, 2011; Verbavatz and Barthelemy, 2020). Thus, the main law of urban population seems to disappear. What, then, of the understanding of urban population dynamics? At this

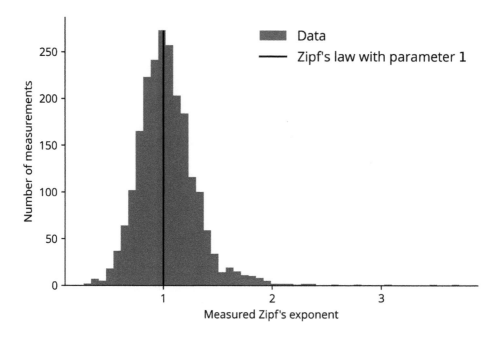

Fig. 3.7: Zipf's exponent distribution for values measured in about 2,000 cases spread over more than 80 studies and collected by Cottineau (2017). Even if the exponent value 1 seems central, empirical estimates can deviate significantly in some countries.

stage, we see that a completely new approach toward urban population statistics is needed. Before reviewing which models have been proposed so far, in the next chapter we will elaborate on the different ways of studying the dynamics of city ranking. Indeed, stating that Zipf's law does not hold does not mean that the very strong hierarchical distribution of cities cannot be investigated in a dynamical manner. On the contrary, from the turbulent dynamics of city hierarchy will emerge new ways of understanding how cities grow.

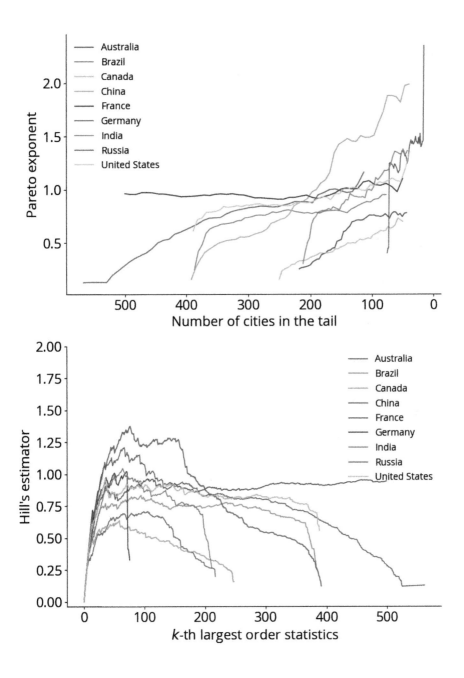

Fig. 3.8: (Top) Value of the exponent obtained by fitting the empirical distribution of city sizes over a set of the n largest cities versus n. (Bottom) Corresponding Hill's plot. Apart from the French case, Zipf's law does not seem to hold in the countries considered here. Data from (Verbavatz and Barthelemy, 2020).

4
Dynamics of ranking

Ranking objects according to a given measure goes far beyond cities, as we can rank virtually anything, from countries, cities and people, to products such as movies or songs (Martínez-Mekler *et al.*, 2009; Blumm *et al.*, 2012; Iñiguez *et al.*, 2022). Given the generality of this problem, we outline in this chapter some theoretical results on the ranking dynamics of any system. The application of these studies to the case of cities certainly requires more work, but the models presented here are interesting starting points for further studies of the ranking dynamics of cities. In addition, the study of city ranking shows that cities exhibit a very turbulent ranking dynamics. This idea will be paramount in future chapters as we attempt to find a new model of city growth.

Stable versus unstable ranking

Empirical analysis

Here we follow the analysis presented in (Blumm *et al.*, 2012). Consider a list of N items with some numerical score $X_i(t)$ which determines their ranking. The highest score corresponds to the first rank $X_i = \max\{X_j\} \Rightarrow r_i = 1$ and the smallest score to the last rank $r = N$. The total score can vary, and it is natural to consider the normalized score (Blumm *et al.*, 2012):

$$x_i(t) = \frac{X_i(t)}{\sum_j X_j(t)}. \tag{4.1}$$

In (Blumm *et al.*, 2012), the authors considered various examples, such as the number of times individual words are used in published journals in a year, the daily market capitalization of companies, the number of diagnoses of a particular disease, the number of annual citations each article receives in the *Physical Review* corpus, and so on. In general, the distribution of x follows a broad distribution covering many orders of magnitude. In some cases, we observe a clear power-law behavior, while for others, there is a cut-off for large x. The dynamics may also be different, with some cases showing a stationary law and others showing temporal variations. Despite these differences, there are some "universal" characteristics. For example, the dispersion of the variation of an item value in a single time step Δx as a function of x generally follows the behavior

$$\sigma_{\Delta x | x} \sim x^\beta, \tag{4.2}$$

where β is in general less than 1 (see Fig. 4.1). This indicates that relative changes (measured by $\sigma_{\Delta x | x}/x$) decrease with x and shows that top-ranked items display relatively smaller changes, a fact also identified in the economic context (Mantegna and Stanley, 1995). In some cases (word usage Medicare, and market capitalization: Fig. 4.1(e)), the score x fluctuates symmetrically around 0, implying that the probability of a score moving up or down is of the same order. This is in contrast with other cases, such as citations, Wikipedia or Twitter (see Fig. 4.1(f)), where fluctuations are asymmetric for large values, showing that a high-scoring item has an increasing tendency to drop (Blumm *et al.*, 2012).

Noise-induced transition

In (Blumm *et al.*, 2012), the authors assumed that the score of an item i follows a simplified Langevin equation of the form already discussed in (Bouchaud and Mézard, 2000):

$$\frac{dx_i}{dt} = f(x_i) + g(x_i)\xi_i(t) - \phi(t)x_i, \tag{4.3}$$

where $f(x)$ represents the deterministic process that governs the score and captures many attributes such as the impact of research papers, the utility of a word, and so on. The multiplicative noise term of the form $g(x_i)\xi_i(t)$ captures the inherent randomness in the system. The noise $\xi_i(t)$ is assumed to be Gaussian, with $\langle\xi_i(t)\rangle = 0$ and $\langle\xi(t)\xi(t')\rangle = \delta(t - t')$, and the amplitude $g(x_i)$ can in general depend on the score x_i. Finally, the last term, $\phi(t)x_i$, ensures that the scores are normalized, so that $\sum_i x_i(t) = 1$. If we sum Eq. 4.3 over i, we then find

$$\sum_i \dot{x}_i = 0 = \sum_i f(x_i) + \sum_i g(x_i)\xi_i(t) - \phi(t)\sum_i x_i, \tag{4.4}$$

which implies that

$$\phi(t) = \sum_i [f(x_i) + g(x_i)\xi(t)] \tag{4.5}$$

$$= \phi_0 + \eta(t), \tag{4.6}$$

where the constant is $\phi_0 = \sum_i f(x_i)$ and the global noise term is $\eta(t) = \sum_i g(x_i)\xi_i(t)$.

Equation 4.3 is obviously too general, and the empirical data suggest some simplifications. First, the drift term $f(x)$ can be written as

$$f(x_i) = A_i x_i^\alpha, \tag{4.7}$$

where it is assumed that the exponent $0 < \alpha < 1$ is the same for all i. The prefactor A_i can be interpreted as the fitness of item i and captures its ability to increase its share x_i (the larger A_i and the larger the growth rate $\dot{x}_i$). The authors of (Blumm *et al.*, 2012) also assume that the noise amplitude behaves as

$$g(x_i) = Bx_i^\beta, \tag{4.8}$$

which is suggested by the empirical measurements on the variance shown in Fig. 4.1c,d (indeed, Eq. 4.3 suggests that $\sigma^2_{\Delta x_i|x_i} \simeq g(x_i)^2\Delta t$). Empirically, the exponent β seems to be comparable for all systems, but the amplitude B displays significant differences

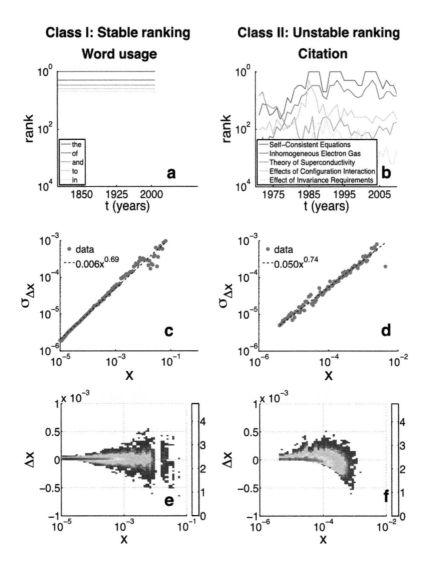

Fig. 4.1: Empirical characteristics of ranking dynamics. On the left, we show an example of stable ranking (word usage) and on the right column an example of unstable ranking (citations). In (a), (b) the evolution of the ranks of five top items in the two systems is shown: word usage displays ranking stability, and citation shows significant volatility. In (c), (d) $\sigma_{\Delta x|x}$ is shown as a function of x and displays a power-law behavior with exponent less than 1. (e),(f) Surface plot of Δx as a function of x. These plots show that for word usage the quantity Δx fluctuates symmetrically around 0, while for citation the fluctuations are asymmetric for large x and indicates that high scoring items have a tendency to drop in score. Figure from (Blumm *et al.*, 2012).

from $B \sim 10^{-3}$ to 10^{-1} (Blumm *et al.*, 2012). This coefficient B is a measure of the noise magnitude, and we expect that the stability of the system will be affected by its value. Indeed, if we denote by $P(x_i, t|A_i)$ the probability that an item i has score x_i at time t given its fitness A_i, its temporal evolution is governed by the Fokker–Planck equation (for the Itô convention, see Chapter 5)

$$\frac{\partial P}{\partial t} = -\frac{\partial}{\partial x_i}[(A_i x_i^\alpha - \phi(t)x_i)P] + \frac{1}{2}\frac{\partial^2}{\partial x_i^2}(B^2 x_i^{2\beta} P). \tag{4.9}$$

This equation cannot be solved in the general case, but if we neglect fluctuations of $\phi(t) \simeq \phi_0$, the time-independent steady-state solution $P_0(x_i|A_i)$ of Eq. 4.9 reads for $\alpha < 1$ (up to a normalization constant)

$$P_0(x_i|A_i) \propto x_i^{-2\beta} e^{\frac{2A_i}{B^2}\frac{x_i^\delta}{\delta}\left[1-\left(\frac{x_i}{x_c}\right)^{1/\gamma}\right]}, \tag{4.10}$$

where $\delta = 1 + \alpha - 2\beta$, $\gamma = 1/(1-\alpha)$ and

$$x_c = \left(\frac{A_i}{\phi_0}\right)\left(\frac{\delta + 1/\gamma}{\delta}\right)^\gamma. \tag{4.11}$$

We introduce $\Psi = \log P_0$ and the most probable value for x_i satisfies $dP_0/dx_i = 0$ implying $d\Psi/dx_i = 0$, which reads, using Eq. 4.10,

$$\Psi'(x_i) \propto A_i x_i^\alpha - \phi_0 x_i - B^2\beta x_i^{2\beta-1} = 0. \tag{4.12}$$

For $B = 0$, we obtain $x_i = 0$ or

$$x_i^* = \left(\frac{A_i}{\phi_0}\right)^\gamma, \tag{4.13}$$

and the condition $\sum_i x_i^* = 1$ then gives $\phi_0 = (\sum_i A_i^\gamma)^{1/\gamma}$.

The stability of the solution is given by the sign of $\Psi''(x_i)$, and in this case it is easy to show that the non-zero solution is stable for $A_i > 0$ and $\alpha < 1$ ($\Psi'' > 0$). When there is noise ($B \neq 0$), it shifts the deterministic solution x_i^* to a new value. If the noise is not too strong (for $B < B_c$), there is a non-zero stable solution and the solution $x_i = 0$, which is unstable. In this low-noise case, the score of an item i will be localized around a value given by the interplay between its fitness and the noise. At $B = B_c$ the non-zero solution disappears, and for $B > B_c$, the distribution behaves as $P(x_i|A_i) \sim x_i^{-2\beta}$ and x_i has large fluctuations. In this case, the value of x_i can have values very different from x_i^* and can vary over many orders of magnitude.

This last discussion is mainly about the value of the score x_i, but doesn't imply the knowledge of the rank of item i. The rank indeed depends on the values of the other items of the systems and, as such, is a collective measure: the rank of i depends on the score x_i and also on the scores of all the other items j. An item can be score-stable with small fluctuations around x_i^* but large enough so that items with similar x^* can swap ranks. The score and rank stability can occur only in the small fluctuations regime $B < B_c$ and the rank stability condition reads

$$\langle x \rangle_r - \langle x \rangle_{r+1} > \sigma_r, \tag{4.14}$$

where σ_r denotes the rank fluctuations of an item at rank r and $\langle x \rangle_r$ denotes the average value of the score at rank r. This condition predicts a second critical value B_r of the noise coefficient. The resulting stability diagram is shown in Fig. 4.2.

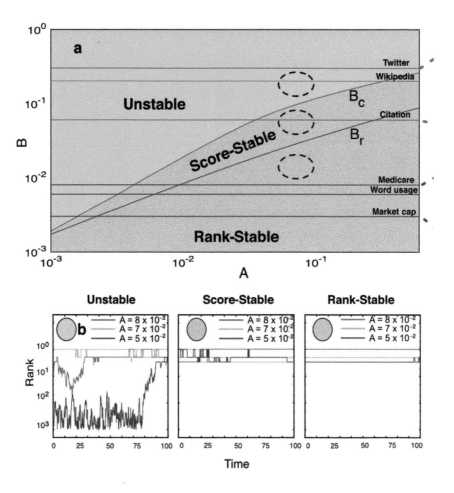

Fig. 4.2: Ranking stability: (a) Phase diagram computed numerically. (b) Time evolution of the ranks for items in each of the three predicted phases (only top-fitness items are shown here). Figure from (Blumm *et al.*, 2012).

The top $B > B_c$ corresponds to the unstable region where the score is broadly distributed and neither score nor rank stability is possible. In the region $B < B_c$, we observe score stability but, as discussed above, this is not enough to ensure rank stability. In the region $B_r < B < B_c$, the score is stable but not the rank, and each item has a score that fluctuates around its steady state x^* determined by its intrinsic fitness. For lower noise $B < B_r$, both scores and ranks are stable. These predictions are illustrated by the temporal profile of ranks shown in Fig. 4.2b in each case.

This simple model thus allows us to discuss the variety of behaviors resulting from the interplay between the intrinsic fitness of an item and the noise existing in the system. In the context of cities, we could well imagine that the fitness A_i can be very different, leading to various simultaneous behaviors, as observed in (Batty, 2006). For large cities, the fitness could be very large and most cities would then be in the rank-stable region. For smaller cities with a smaller fitness, depending on the noise, we could be in different regions of the phase diagram Fig. 4.2a. Of course, this model is not justified from basic principles and should probably not be trusted in its details, but it has the merit of connecting fluctuations of scores and ranks. In this respect, it is a very useful guide for understanding the rank dynamics. There is, however, an important ingredient that this phenomenological approach misses: the number of items is not fixed. For cities, new cities can emerge and others can disappear, and there is a total net flux of new elements in the ranking list. This ingredient seems to be crucial and will be discussed in the next section.

Modeling the ranking dynamics

Open versus closed systems

Ranked lists, such as the top 100 universities or the Fortune 500 companies, typically have a fixed size N_0. The different items can, however, enter or leave the list at any time of the observation period (denoted here by $t = 0, 1, \ldots, T - 2$). For given values of N_0 and T, it is then possible to introduce two measures of the flux (Iñiguez *et al.*, 2022). The first one is the *rank turnover*,

$$o_t = \frac{N_t}{N_0}, \tag{4.15}$$

where N_t is the number of distinct items seen in the ranking list until time t. The instantaneous turnover rate is $(N_{t+1} - N_t)/N_0$, and on average over the whole period this is given by

$$\dot{o} = \frac{o_{T-1} - o_0}{T - 1}. \tag{4.16}$$

The second measure discussed in (Iñiguez *et al.*, 2022) is the *rank flux* F_t, which measures the probability that an item enters or leaves the ranking list at time t. The average turnover rate and the flux averaged over time $F = \langle F_t \rangle$ are in general highly correlated variables, as we can observe in Fig. 4.3.

In the study (Iñiguez *et al.*, 2022), the authors distinguish two types of systems according to their values of F and $\dot{o}$:

- Open systems with $F \sim \dot{o} \sim 1$ have items that constantly enter and leave the list. The ranking list thus displays significant variations in time. Typically, the ranking of universities is one such open system. In these systems, only the top is

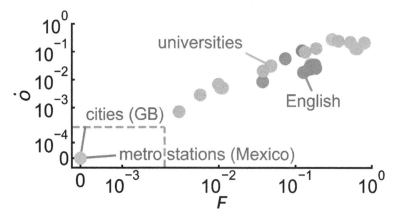

Fig. 4.3: Average turnover rate $\dot{o}$ versus average flux F for various systems. We observe in general a strong correlation between these two measures. For the set of cities considered in (Iñiguez *et al.*, 2022), F and $\dot{o}$ are both very small and the system is essentially closed with a fixed list of cities. Figure from (Iñiguez *et al.*, 2022).

stable, and as we go down in the list, rank trajectories increasingly fluctuate in time.

- Almost closed systems $F \sim \dot{o} \sim 0$ have small variations in their ranking list contents. Completely closed systems can also be found with $F = \dot{o} = 0$ and where no single new item is recorded during the observation window. In this case, low-ranked elements are also stable. It should be noted that in the study (Iñiguez *et al.*, 2022), UK cities in the twentieth century were considered and displayed a strong stability. Obviously other systems—such as US cities from 1790 to nowadays—would certainly display a larger turnover rate and flow. The classification as open or closed systems depends, therefore, on the observation period and the development stage of the system studied. We also note here that the size of the ranking list certainly has an impact that should be taken into account.

These results are confirmed by measures of the rank change C, defined as the average probability that an item at rank R changes between time $t-1$ and t. As expected, we observe that for open systems, C is essentially an increasing function of R and for less open systems, C has a symmetric shape, indicating that mid-range ranks are the most likely to vary.

When the rank flux F is low, both the top and the bottom of the rank list are stable and the dynamics result from an interplay between replacement and displacement. When systems become more open, there is a growing flux of items at the bottom of the ranking list. This breaks the symmetry and C becomes on average an increasing function of R. Obviously a more detailed inspection of the impact of the rank list size N_0 on the various results is needed.

A simple model of rank dynamics

In order to understand certain aspects of the dynamics of ranks in various systems, (Iñiguez *et al.*, 2022) proposed a simplified model that implements only two mechanisms (see Fig. 4.4):

1. Random displacements of elements across the list
2. Random replacements of elements by new ones.

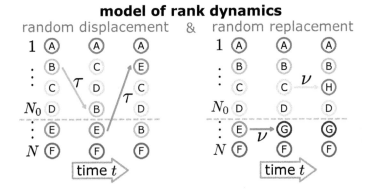

Fig. 4.4: Model of rank dynamics proposed in Iñiguez *et al.* (2022). The system is made of N elements and the ranking list size is N_0. At each time step, an item chosen at random is moved to a random rank with probability τ. A random item is also replaced by a new item with probability ν. Figure from (Iñiguez *et al.*, 2022).

The length of the list is N_0 and the total number of available items is $N > N_0$. The ratio is denoted by $p = N_0/N \leq 1$. It is assumed that an action (replacement or displacement) takes place every $\Delta t = 1/N$, so that between t and $t+1$, N modifications took place in the list (the time step varies from 0 to $N(T-1)$ and time varies from $t = s/N = 0$ to $T-1$). The normalized rank $r = R/N$ varies, then, between $1/N$ and 1, and what happens for $r > p$ is ignored (i.e. for items out of the list). At each time step s, two independent rank updates are performed: first, a random displacement with probability τ and second, a replacement with probability ν. For the random displacement, an item is selected at random, removed and replaced at random in the list of N elements. For the replacement with probability ν, an item is chosen at random in $[\![0, N]\!]$ and replaced by a new item from outside the system, leaving the rest untouched. Empirical data fix the values of T and N_0 and there are three parameters left: τ, ν and p. For closed systems, we can take $N = N_0$ and $\nu = 0$, and the dynamics is driven by diffusion and controlled by one free parameter, τ. For open systems, it is natural to choose $N = N_{T-1}$, and the free parameters are then τ and ν.

The temporal evolution of this model can be described by the probability $P(r \to r', t)$ to move from the normalized rank $r = R/N$ to the normalized rank $r' = r/N$ after a time t. For a single time step $\Delta r = 1/N$, an item that is not replaced (which happens with probability $1 - \nu\Delta r$) moves from r to r' with a probability τ, either by "direct" displacement with probability L or due to the rank change of another element (with probability $D(r, r')$), or stays in place with probability $1 - \tau$. We can then write

$$P(r \rightarrow r', \Delta r) = (1 - \nu \Delta r)[\tau(L + D(r,'r)) + (1 - \tau)\delta_{r,r'}], \qquad (4.17)$$

where $\delta_{r,r'}$ is the Kronecker delta. For a uniform process and for $r = \Delta r, 2\Delta r, \dots, 1$, the probability L to move from r to r' is simply given by $L = \frac{1}{N} \times \frac{1}{N} = \Delta r^2$.

The quantity $D(r, r')$ is more complicated to compute. In a single step, this quantity is non-zero only for neighbors $r = r' \pm \Delta r$ or for $r' = r$. It is then easy to show that

$$D(r, r') = \begin{cases} r(1 - r) & \text{for } r = r' - \Delta r \\ (r - \Delta r)^2 + (1 - r)^2 & \text{for } r = r' \\ (r - \Delta r)(1 - r + \Delta r) & \text{for } r = r' + \Delta r \\ 0 & \text{otherwise.} \end{cases} \qquad (4.18)$$

Using these results, one can write a master equation for $P(r \rightarrow r', t)$ (see the supplementary information in (Iñiguez *et al.*, 2022) for details), and finally gets

$$P(r \rightarrow r', t) = e^{-\nu t}(L_t + D(r, r', t)), \qquad (4.19)$$

where $L_t = (1 - e^{-\tau t})/N$ is the probability that an item is selected and jumps to any other rank in time t. The probability $D(r, r', t)$ that an element in rank r gets displaced to r' in time t is approximatively given by the following equation written in the continuous limit:

$$\frac{\partial D}{\partial t} = \alpha r(1 - r)\frac{\partial^2 D}{\partial R^2}, \qquad (4.20)$$

where $\alpha = \tau/N$. This is the Wright–Fisher equation (see (Iñiguez *et al.*, 2022) and references therein), and its solution $D(r, t)$ can be approximated by a Gaussian function whose average is the initial position r_0 and whose standard deviation is given by $\sigma_t = \sqrt{2\alpha r_0(1 - r_0)t}$

$$D(r, t) \simeq e^{-\tau t}\frac{1}{\sigma_t\sqrt{2\pi}}e^{-\frac{(r-r_0)^2}{2\sigma_t^2}}. \qquad (4.21)$$

One then obtains (see Supplementary Information of (Iñiguez *et al.*, 2022) for details)

$$P(r_0 \rightarrow r, t) \simeq e^{-\nu t}\left[\Delta r_0(1 - e^{-\tau t}) + \sqrt{\frac{\Delta r}{4\pi\tau r_0(1 - r_0)t}}e^{-\frac{(r-r_0)^2}{4\tau r_0(1 - r_0)t\Delta r} - \tau t}\right]. \qquad (4.22)$$

It is also possible to derive expressions (in the large t limit) for the mean flux given by

$$F = 1 - \sum_{x=\Delta r}^{p} P(r_0 \rightarrow r, t) \qquad (4.23)$$

$$= 1 - e^{-\nu}\left(p + (1 - p)e^{-\tau}\right) \qquad (4.24)$$

and the turnover rate $\dot{o}$ (Iñiguez *et al.*, 2022)

$$\dot{o} = \nu \frac{\nu + \tau}{\nu + p\tau}.\tag{4.25}$$

From the data, we can thus measure F and $\dot{o}$ (and p in general estimated from N_{T-1}), and numerically inverting these equations allows us to determine the parameters ν and τ.

Rank variations of cities

Considering the rank variation of cities within a defined system of cities (typically a bounded territory or a country) and over a large period of time amounts to comparing how cities evolve and lays the groundwork for explaining how historical changes affect the spatial distribution of populations. In (Batty, 2006), the geographer Batty studied the population dynamics of major cities in the US since 1790, in the UK since 1901 and in world cities from 430 BCE onward, and compared how they evolved over time. To measure and interpret rank variations, Batty used both quantitative and graphical descriptions of the rank dynamics. Defining $r_i(t)$ as the rank of city i at time t, he first studied the *average rank variation per unit of time*,

$$d = \frac{1}{NT} \sum_t \sum_{i=1}^{N} (r_i(t) - r_i(t-1)),\tag{4.26}$$

during T years and for N cities, which he showed to be different from what could have been expected from Gabaix's model (see Chapter 6).

Batty also plotted the rank variation of cities within a system on a *rank clock*, showing temporal trajectories of each city in rank-size space. These clocks allow cities to be ranked by population on a clock face, with the largest city on the outside (of rank $r = 1$) and the rank decreasing as one approaches the centre (the other convention with rank increasing from the center is also used). Batty then showed that city dynamics are very turbulent, with large cities appearing or disappearing on very short scales. In the spirit of (Batty, 2006), we show in Fig. 4.5 how such rank clocks can help fathom the dynamics of representative behaviors of French cities.

Following (Iñiguez *et al.*, 2022), we also computed the turnover rate $\dot{o}$ and the average flux F defined above in two different historical systems of cities (American largest cities between 1790 and 1990 and French cities between 1876 and 2015). The computed values are given in Table 4.1. As expected, we see that the dynamics of US cities exhibit a large turnover rate and are hence a more open system than the system of French cities.

In this chapter, we have seen that although further research is needed in order to have an exhaustive understanding of the ranking dynamics of cities, these dynamics are very turbulent with, in general, a large turnover. As shown by (Batty, 2006), the dynamics of cities is not compatible with the usual models of city growth. In the next part, we will review these different models and their historical links, and we will point to their main limitations. This review work will be indispensable when we come to propose a new model of city growth in Chapter 9.

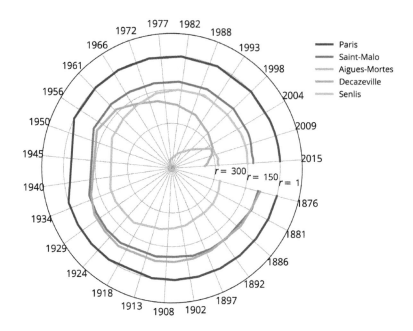

Fig. 4.5: Changes in the population rank of several French cities over time. While the trajectories of cities such as Paris, Saint-Malo and Senlis are relatively stable, the examples of Aigues-Mortes and Decazeville show that the destiny of a city can change in very short time scales. Thus, the rank order of cities within a closed system can be very turbulent, as highlighted in (Batty, 2006).

Table 4.1: Average rank variation per unit of time d as defined in (Batty, 2006), turnover rate $\dot{o}$ and average rank flux F defined in (Iñiguez *et al.*, 2022) in two systems: US cities between 1790 and 1990 and French cities between 1876 and 2015. In the US, we study the rank variations among the $N_0 = 20$ largest cities for each census year. Over time, $N_{T-1} = 60$ cities have been among the 20 largest cities for at least one year, showing that the system of US cities exhibits a high turnover rate and rank flux. On the contrary, rank variations in France have occurred in a more closed system, with only $N_{T-1} = 700$ cities being among at least one census year in the Top 500 French cities. Reducing the rank list to 20 cities in France to make comparison with the US more robust does not change the quantitative difference in the values of the turnover rate and the rank flux. The system of French cities has small values and is almost closed (in all cases, a more thorough discussion about the impact of the ranking list size is certainly needed).

System	d	$p = N_0/N_{T-1}$	$\dot{o}$	F
Top 20 US cities 1790–1990	4.7	20/60	0.10	0.23
Top 500 French cities 1876–2015	6.0	500/700	0.018	0.086
Top 20 French cities 1876–2015	/	20/28	0.014	0.039

Models of urban growth

5
Stochastic calculus

To understand the modeling of city growth, we need to resort to the mathematical formalism of stochastic differential equations (SDEs). SDEs are dynamical equations in which future events are not purely deterministic; they depend on some noise that happens at each time. The noise term models everything that is too complex to be written explicitly: particle shocks in Brownian motion, bid and ask in stock market prices, migrations, births and deaths in city growth. This is the main reason why this type of equation appears in so many different fields of research and is a fundamental tool in the study of complex systems.

Historically, the first modeling of such processes was Brownian motion (see, e.g., (Gardiner, 1985)). Interestingly, its mathematical description occurred almost simultaneously in social and natural sciences. In 1901, Louis Bachelier proposed a mathematical model of Brownian motion applied to finance, while Albert Einstein developed a model for physics in 1905. The earliest example of a stochastic differential equation is called the Langevin equation (after the French physicist) and describes the motion of an harmonic oscillator subject to a random force. From a mathematical point of view, these objects are not always well defined, and their theory was the subject of intense research. In particular, the Japanese mathematician Kiyoshi Itô introduced the idea of stochastic integrals, while a parallel but different approach was developed by the Russian physicist Ruslan Stratonovich. Both these scientists contributed to identifying and solving mathematical ambiguities that can appear in the case of SDEs with multiplicative noise.

While the number of books on the subject is substantial, we propose here a very short introduction to these stochastic differential equations. Starting with the notion of the random walk, we will show how to use these nontrivial objects and will avoid the use of complex mathematical objects. As we will see, caution is required in using SDEs, as different definitions of integration lead to different results. Finally, we will elaborate on the particular case of multiplicative noise that appears to be fundamental for understanding many systems and in particular the growth of cities.

Brownian motion

The discrete random walk

We consider the simple stochastic process where a particle moves on a one-dimensional lattice of spacing a according to the result of tossing a coin at each time step. For each tail, the particle moves one step to the right, and for each head, one step to the left. If heads and tails are equiprobable, the position $x(t)$ of the particle at time t is

$$x(n) = \sum_{k=1}^{n} \xi(k) \quad \text{where} \quad \xi(k) = \begin{cases} a & \text{with prob} \quad 1/2 \\ -a & \text{with prob} \quad 1/2. \end{cases} \quad (5.1)$$

If coin tosses are independent, the process is Markovian and one says that the particle follows a discrete symmetric random walk. The position at time t is exactly described by the probability distribution

$$\mathbb{P}(x(n) = ka) = \binom{n}{\frac{n+k}{2}} \frac{1}{2^t} \quad \text{for} \quad k \in \{-n, -n+2, \dots, n-2, n\}. \quad (5.2)$$

where the binomial coefficient reads $\binom{n}{k} = \frac{n!}{k!(n-k)!}$. The average position is $\langle x(n) \rangle = 0$ and the variance is $\langle x^2(n) \rangle = na$.

When $n \to \infty$, it results from the central-limit theorem (or alternatively from Stirling's formula) that the probability distribution of the renormalized particle position is well approximated by a normal law,

$$\frac{x(n)}{\sqrt{n}} \sim \mathcal{N}(0, a). \quad (5.3)$$

In the large n limit, the walk is diffusive and approximates a Brownian motion that we will discuss in the next section.

Wiener process

We consider a time interval t subdivided into n equal intervals of duration $\tau = t/n$ and a walk process similar to the previous one of the form

$$x(t) = \sum_{k=1}^{n} \xi(k). \quad (5.4)$$

This time the process is defined on a continuous space where each move $\xi(k)$ follows a Gaussian law of variance τ:

$$\xi(k) \sim \mathcal{N}(0, \tau). \quad (5.5)$$

In the scaling limit where $\tau \to 0$ and $n \to \infty$ while keeping $t = n\tau$ fixed, we say that the particle obeys a Brownian motion, and the following quantity is called a Wiener process:

$$x(t) = \lim_{\tau \to 0} \sum_{i=1}^{n=t/\tau} \xi(i) \equiv \int_0^t d\xi(t'), \quad (5.6)$$

where $d\xi(t)$ represents the infinitesimal random move at time t. By positing $dt \equiv \tau$, we note that the order of magnitude of $d\xi(t)$ is not dt but $\sqrt{dt}$.

More formally, the Wiener process verifies the following properties:

1. $x(0) = 0$.
2. $\forall t > 0$, $\forall u \geq 0$, $x(t+u) - x(t)$ is independent of $x(s)$ for $s \leq t$.
3. $\forall t, u > 0$, $x(t+u) - x(t) \sim \mathcal{N}(0, u)$.

4. $x(t)$ is continuous in time.

In particular, we note that $x(t) \sim \mathcal{N}(0, t)$ and $\langle x(t_1)x(t_2)\rangle = \min(t_1, t_2)$.

One can extend this elementary Wiener process to the Wiener process with drift defined by

$$x(t) = \mu t + \sigma W(t), \tag{5.7}$$

where $W(t)$ is a Wiener process. It can be rewritten into the form of a stochastic differential equation for the particle position $x(t)$ as

$$\frac{\mathrm{d}x(t)}{\mathrm{d}t} = \mu + \sigma \frac{\mathrm{d}W(t)}{\mathrm{d}t} \equiv \mu + \sigma \eta(t). \tag{5.8}$$

The noise $\eta(t)$, which is the "derivative of the Wiener process," is called the Langevin noise. Since $\eta(t) = \frac{x(t+\mathrm{d}t)-x(t)}{\mathrm{d}t}$ is a linear combination of Gaussian processes, it is a Gaussian process characterized by its first two moments:

$$\langle \eta(t)\rangle = 0 \quad \text{and} \quad \langle \eta(t)\eta(t')\rangle = \delta(t - t'), \tag{5.9}$$

where δ is the Dirac delta function. The presence of the delta function in the correlations expresses the fact that the process is Markovian and hence that moves at time t and t' are independent. By analogy with signal theory, it is called a white noise, since the Fourier transform of the correlation function is flat.

The noise $\eta(t)$ is, however, ill-defined: it scales as $1/\sqrt{\mathrm{d}t}$ and diverges when $\mathrm{d}t \to 0$. If the Wiener process is continuous and well-defined, its derivative is ambiguous and requires additional work. The precise sense of the white noise will depend on the way in which the derivatives and integrations are performed.

Itô and Stratonovich prescriptions

Stochastic integrals

We start from a generalized stochastic differential equation which extends the Langevin equation Eq. 5.8 to a general drift $f(x)$ and a prefactor $g(x)$ of the noise

$$\frac{\mathrm{d}x(t)}{\mathrm{d}t} = f(x(t)) + g(x(t))\eta(t). \tag{5.10}$$

The solution of Eq. 5.10 depends on the noise and is therefore stochastic. It can be written as

$$x(t) = \int_0^t \mathrm{d}s \; (f(x(s)) + g(x(s))\eta(s)). \tag{5.11}$$

As stated above, the noise $\eta(t)$ itself is ambiguous, and we need a proper way to define the stochastic integral

$$\int_{t'}^t g(x(s))\eta(s)\mathrm{d}s \equiv \int_{t'}^t g(x(s))\mathrm{d}W, \tag{5.12}$$

where W is the underlying Wiener process. In general, one can evaluate the integral in Eq. 5.12 between time t and $t + h$ as

$$\int_t^{t+h} g(x(s))dW = g\left(\alpha x(t) + (1-\alpha)x(t+h)\right)\left[W(t+h) - W(t)\right] \tag{5.13}$$

when $h \to 0$ and where $\alpha \in [0,1]$. This expression means that the function g is evaluated at some point between $x(t)$ (for $\alpha = 1$) and $x(t+h)$ (for $\alpha = 0$). In the Itô prescription, one chooses $\alpha = 1$: the function g is evaluated at the initial time. Over the interval $[t, t']$ subdivided into N intervals, Itô calculus enables us to compute

$$\int_t^{t'} g(x(s))dW = \lim_{N\to\infty} \sum_{i=1}^{N} g(x(t_{i-1}))\left[W(t_i) - W(t_{i-1})\right], \tag{5.14}$$

where $t_i = t + (t'-t)/N$. Alternatively, the Stratonovich prescription chooses $\alpha = 1/2$: the function g is evaluated at the average point between initial and final times. Over the interval $[t, t']$ subdivided into N intervals, Stratonovich calculus gives then

$$\int_t^{t'} g(x(s))dW = \lim_{N\to\infty} \sum_{i=1}^{N} g\left(\frac{x(t_i) + x(t_{i-1})}{2}\right)\left[W(t_i) - W(t_{i-1})\right], \tag{5.15}$$

where $t_i = t + (t' - t)/N$.

Both prescriptions give a well-defined meaning to the integral in Eq. 5.12 and to the SDE in Eq. 5.10. They define different conventions and each one gives a unique meaning to stochastic integration. This discussion is not formal, and it should be noted that, crucially, these conventions lead to different results, which highlights their importance.

As a practical illustration of these conventions, we now consider the following integral:

$$\int_t^{t'} W(s)dW. \tag{5.16}$$

With the Itô prescription, this expression is defined as

$$\int_t^{t'} W(s)dW = \lim_{N\to\infty} \sum_{i=1}^{N} W(t_{i-1})\left[W(t_i) - W(t_{i-1})\right] \tag{5.17}$$

$$= \frac{W(t')^2 - W(t)^2}{2} - \frac{1}{2}\lim_{N\to\infty}\sum_{i=1}^{N}\left[W(t_i) - W(t_{i-1})\right]^2, \tag{5.18}$$

whereas Stratonovich calculus yields

$$\int_t^{t'} W(s)dW = \lim_{N\to\infty} \sum_{i=1}^{N} \frac{W(t_i) + W(t_{i-1})}{2}\left[W(t_i) - W(t_{i-1})\right] \tag{5.19}$$

$$= \frac{W(t')^2 - W(t)^2}{2}. \tag{5.20}$$

Hence, Stratonovich's prescriptions lead to a result similar to ordinary calculus, whereas an unexpected additional term appears in Itô's prescription. In that sense, Stratonovich's prescription is more natural, but in this convention the position $x(t)$ depends on the noise $\eta(t)$ at time t, whereas they are uncorrelated in Itô's prescription,

$$\langle x(t)\eta(t)\rangle = 0. \tag{5.21}$$

Another way of looking at Itô's prescription is that the function is not anticipating the action of the noise: in the integral $\int g(x(t))dW$ the function is first computed (at $g(x(t))$) and the noise acts "after."

Finally, one can note that both interpretations lead to the same result when the noise is purely additive—that is, when $g(x(t))$ is independent of $x(t)$.

Physical meaning

There is another, more intuitive, way to define Stratonovich's and Itô's prescriptions (Bouchaud and Mézard, 2016). This "physical" approach is equivalent to the one previously studied in (Hairer and Pardoux, 2015). We define the correlator

$$G(t,t') \equiv \langle \eta(t)\eta(t')\rangle, \tag{5.22}$$

and in Eq. 5.9 the noise $\eta(t)$ is simply described as a Gaussian process with zero correlation, where $G(t,t') = \delta(t-t')$.

As a continuous Gaussian process, the probability of a given realization of noises $\{\eta(t)\}$ can be represented by a path integral

$$\mathbb{P}\left(\{\eta(t)\}, 0 \leq t \leq T\right) = \mathcal{N}\mathcal{D}\eta(t) \exp\left(-\frac{1}{2}\int_0^T\int_0^T dtdt'\eta(t)G^{-1}(t,t')\eta(t')\right), \tag{5.23}$$

where $\mathcal{D}\eta(t)$ is an infinite dimensional measure which hides many subtleties but can be seen as the continuous limit of the product of discrete measures:

$$\mathcal{D}\eta(t) = \lim_{N\to\infty}\prod_{i=1}^N d\eta_i(t). \tag{5.24}$$

In the absence of correlations, $G^{-1}(t,t') = G(t,t') = \delta(t-t')$ and

$$\mathbb{P}\left(\{\eta(t)\}, 0 \leq t \leq T\right) = \mathcal{N}'\exp\left(-\frac{1}{2}\int_0^T dt\eta(t)^2\right). \tag{5.25}$$

Eq. 5.23 can, however, be generalized to any correlator. One interesting case is when the correlator is an exponential function of the form

$$G(t,t') = \frac{1}{2\tau_c}\exp\left(-\frac{|t-t'|}{\tau_c}\right), \tag{5.26}$$

where we say that the noise $\eta(t)$ follows an Ornstein–Uhlenbeck process. The inversion of the correlator can be done in Fourier space where

$$G(\omega) = \frac{1}{1+\omega^2\tau_c^2}, \tag{5.27}$$

implying that

$$G^{-1}(\omega) = 1 + \omega^2 \tau_c^2. \tag{5.28}$$

When $\tau_c \to 0$, $G(\omega)$ becomes flat and justifies the notion of white noise: all frequencies appear while the noises at time t and t' become uncorrelated. In the presence of correlations ($\tau_c \neq 0$), the Lorentzian form of Eq. 5.28 implies that all frequencies (colors) are not equiprobable: very high frequencies are truncated and the noise is colored ("red" in this case). In the time representation, the inverse correlator is

$$G^{-1}(t, t') = \delta(t - t') \left(1 + \tau_c^2 \partial_t \partial_{t'}\right) \tag{5.29}$$

and a specific realization of noises has weight

$$\mathbb{P}\left(\{\eta(t)\}, 0 \leq t \leq T\right) = \mathcal{N}' \exp\left(-\frac{1}{2} \int_0^T dt (\eta(t)^2 + \tau_c^2 \dot{\eta}(t))\right). \tag{5.30}$$

The correlation time τ_c in the Ornstein–Uhlenbeck noise prevents the noise pulses from changing too fast and smoothes the process: $\eta(t) \sim \frac{1}{\sqrt{\tau_c}}$ is finite. It thus introduces a second time scale in the system (alongside dt). Having $\tau_c \neq 0$ is more "physical," since infinite noise pulses are impossible in physical situations. The rule of integration of the noise of the Langevin equation $\dot{x}(t) = f(x) + \eta(t)$ will then depend on the relation between the scales dt and τ_c, and we have to distinguish between two cases:

- When $\tau_c \gg dt$, the system is not pathological, $\eta(t)$ follows an Ornstein–Uhlenbeck process and is finite. The rules of ordinary calculus apply to any function g:

$$\frac{dg}{dt} = \frac{\partial g}{\partial t} + \frac{\partial g}{\partial x} \dot{x}. \tag{5.31}$$

  Since $\tau_c \neq 0$, the noise at time t is, however, not uncorrelated from the past positions. This justifies the equivalence between taking Ornstein–Uhlenbeck noises and Stratonovich's prescription of integration. They are, in fact, the same as proven in (Hairer and Pardoux, 2015). Correlations between noise at time t and past positions smooth out the trajectory: rules of ordinary calculus apply at the expense of losing uncorrelated noises (the process is no longer Markovian).

- On the contrary, when $\tau_c \ll dt$, we go back to the problem of an ill-defined noise with $\eta(t) \sim \frac{1}{\sqrt{dt}}$. In order to keep uncorrelated noises and the constraint $\langle g(x(t))\eta \rangle = 0$ for any function g, one has to resort to Itô's rule of integration in which Itô's lemma shows that the correct rule of integration is

$$\frac{dg}{dt} = \frac{\partial g}{\partial t} + \frac{\partial g}{\partial x} \dot{x} + \frac{1}{2} \frac{\partial^2 g}{\partial x^2}, \tag{5.32}$$

  the last term of the right-hand side being Itô's correction (see, e.g., (Gardiner, 1985) for a very pedagogical presentation of Itô's lemma).

To summarize, Stratonovich's rule of integration corresponds to more physical situations where noises are correlated and bounded: classical rules of integration apply at the expense of losing uncorrelated noise. In order to make it possible to preserve an uncorrelated noise, one has to resort to different calculus and add a correction to the derivation rules, as proposed by Itô.

Fokker–Planck equation

So far, we have focused on trajectories. We used stochastic differential equations to describe the evolution of a variable. We saw that these equations are ill-defined and that we have to choose a prescription (either Itô's or Stratonovich's). An alternative way of looking at SDEs is to focus on probabilities rather than trajectories. In this case, we are interested in finding the equation that governs the evolution of the probability $p(x, t)$ that the variable has the value x at time t. This is the Fokker–Planck equation (sometimes called the Kolmogorov forward equation) associated with the SDE under study.

We illustrate this discussion on the following SDE:

$$\frac{\mathrm{d}x}{\mathrm{d}t} = f(x(t)) + g(x(t))\eta(t), \tag{5.33}$$

where $\eta(t)$ is a Gaussian white noise of zero average $\langle \eta(t) \rangle = 0$ and correlation function $\langle \eta(t)\eta(t') \rangle = \delta(t - t')$. We will derive the corresponding Fokker–Planck equation with heuristic arguments taken from (Hurtado, 2023).

We introduce the time discretization $t_i = t_0 + ih$ with $h \ll 1$ and $x_i = x(t_i)$ and the evolution equation (Eq. 5.33) can be rewritten as the following recurrence:

$$x_{i+1} = x_i + \int_{t_i}^{t_{i+1}} \mathrm{d}s \ (f(x(s)) + g(x(s))\eta(s)). \tag{5.34}$$

As we saw in Eq. 5.13, we use a discretization method with weight α between positions i and $i + 1$ for the random term, and writing $dW = \eta ds$

$$\int_{t_i}^{t_{i+1}} \mathrm{d}s \ g(x(s))\eta(s) \simeq g\left(\alpha x_i + (1 - \alpha)x_{i+1}\right) \left(W(t_{i+1}) - W(t_i)\right)$$

$$= \sqrt{h} u_i g\left(\alpha x_i + (1 - \alpha)x_{i+1}\right), \tag{5.35}$$

where we have used the property of the Wiener process $W(t_{i+1}) - W(t_i) \sim \sqrt{h} u_i$ where u_i is a Gaussian random variable of zero mean $\langle u_i \rangle = 0$, and variance given by $\langle u_i u_j \rangle = \delta_{ij}$.

For consistency, we also write

$$\int_{t_i}^{t_{i+1}} \mathrm{d}s \ f(x(s)) \simeq hf(\alpha x_i + (1 - \alpha)x_{i+1}) \tag{5.36}$$

so that

$$x_{i+1} = x_i + hf_\alpha^i + \sqrt{h} u_i g_\alpha^i, \tag{5.37}$$

with $f_\alpha^i = f(\alpha x_i + (1 - \alpha)x_{i+1})$ and $g_\alpha^i = g(\alpha x_i + (1 - \alpha)x_{i+1})$.

Since $h \ll 1$, we can use Taylor expansions of the functions

$$f_\alpha^i = f\left(x_i + (1 - \alpha)(hf_\alpha^i + \sqrt{h} u_i g_\alpha^i)\right)$$

$$= f(x_i) + O(\sqrt{h}) \tag{5.38}$$

and

$$g_\alpha^i = g\left(x_i + (1-\alpha)(hf_\alpha^i + \sqrt{h}u_i g_\alpha^i)\right)$$
$$= g(x_i) + (1-\alpha)\sqrt{h}u_i g_\alpha^i g'(x_i) + O(h)$$
$$= g(x_i) + (1-\alpha)\sqrt{h}u_i g(x_i)g'(x_i) + O(h). \tag{5.39}$$

Collating all these elements together leads to

$$x_{i+1} = x_i + \sqrt{h}u_i g(x_i) + h\left(f(x_i) + (1-\alpha)u_i^2 g(x_i)g'(x_i)\right) + O(h^{3/2}). \tag{5.40}$$

We define now the probability $p(x_i, t_i)$ of being at position x_i at time t_i. Using the previous equation, we obtain the recurrence relation between $p(x_{i+1}, t_{i+1})$ and all the possible positions x_i at time t_i:

$$p(x_{i+1}, t_{i+1}) =$$
$$\left\langle \int dx_i\, p(x_i, t_i)\delta\left(x_i + \sqrt{h}u_i g(x_i) + h\left(f(x_i) + (1-\alpha)u_i^2 g(x_i)g'(x_i)\right) - x_{i+1}\right)\right\rangle_{u_i},$$
$$\tag{5.41}$$

where the brackets $\langle\cdot\rangle_{u_i}$ represent the average over all possible noises u_i and the delta function ensures that x_{i+1} is given by Eq. 5.40. We now expand the delta function for $h \ll 1$ and obtain

$$\delta\left[x_i + \sqrt{h}u_i g(x_i) + h\left(f(x_i) + (1-\alpha)u_i^2 g(x_i)g'(x_i)\right) - x_{i+1}\right] = \delta(x_i - x_{i+1})$$
$$+ \left(\sqrt{h}u_i g(x_i) + h\left(f(x_i) + (1-\alpha)u_i^2 g(x_i)g'(x_i)\right)\right)\frac{\partial}{\partial x_i}\delta(x_i - x_{i+1})$$
$$+ \frac{1}{2}hu_i^2 g^2(x_i)\frac{\partial^2}{\partial x_i^2}\delta(x_i - x_{i+1}) + O\left(h^{3/2}\right). \tag{5.42}$$

We note here that x_{i+1} is exogenous in the calculation, so that for any function h, $\langle h(x_{i+1})u_i\rangle_{u_i} = 0$ and $\langle h(x_{i+1})u_i^2\rangle_{u_i} = h(x_{i+1})$. Using this expansion in Eq. 5.41, we obtain

$$p(x_{i+1}, t_{i+1}) = \overbrace{\left\langle \int dx_i\, p(x_i, t_i)\delta(x_i - x_{i+1})\right\rangle_{u_i}}^{p(x_{i+1}, t_i)}$$

$$+ \sqrt{h}\overbrace{\left\langle \int dx_i\, p(x_i, t_i)u_i g(x_i)\frac{\partial}{\partial x_i}\delta(x_i - x_{i+1})\right\rangle_{u_i}}^{0 \text{ since } \langle u_i\rangle = 0}$$

$$+ h\left\langle \int dx_i\, p(x_i, t_i)f(x_i)\frac{\partial}{\partial x_i}\delta(x_i - x_{i+1})\right\rangle_{u_i}$$

$$+ h(1-\alpha)\left\langle \int dx_i\, p(x_i, t_i)u_i^2 g(x_i)g'(x_i)\frac{\partial}{\partial x_i}\delta(x_i - x_{i+1})\right\rangle_{u_i}$$

$$+ \frac{h}{2}\left\langle \int dx_i\, p(x_i, t_i)u_i^2 g^2(x_i)\frac{\partial^2}{\partial x_i^2}\delta(x_i - x_{i+1}) + O\left(h^{3/2}\right)\right\rangle_{u_i}. \tag{5.43}$$

We now use integration by parts and the relations $\langle u_i \rangle = 0$ and $\langle u_i^2 \rangle = 1$, yielding

$$p(x_{i+1}, t_{i+1}) = p(x_{i+1}, t_i) + h \left(-\frac{\partial}{\partial x_{i+1}} f(x_{i+1})p(x_{i+1}, t_i) \right.$$
$$- (1 - \alpha)\frac{\partial}{\partial x_{i+1}} g(x_{i+1})g'(x_{i+1})p(x_{i+1}, t_i)$$
$$\left. + \frac{1}{2}\frac{\partial^2}{\partial x_{i+1}^2} g(x_{i+1})p(x_{i+1}, t_i) \right) \tag{5.44}$$

In the limit $h \to 0$ limit, we finally obtain the Fokker–Planck equation (FPE) associated with Eq. 5.33:

$$\frac{\partial}{\partial t}p(x, t) = -\frac{\partial}{\partial x}\left[(f(x) + (1 - \alpha)g(x)g'(x))\, p(x, t) \right] + \frac{1}{2}\frac{\partial}{\partial x^2}\left[g(x)^2 p(x, t) \right], \tag{5.45}$$

where α determines the prescription (Itô or Stratonovich). If we choose $\alpha = 1$, we obtain Itô's FPE:

$$\frac{\partial}{\partial t}p(x, t) = -\frac{\partial}{\partial x}[f(x)p(x, t)] + \frac{1}{2}\frac{\partial}{\partial x^2}[g(x)^2 p(x, t)], \tag{5.46}$$

and $\alpha = 1/2$ leads to Stratonovich's FPE:

$$\frac{\partial}{\partial t}p(x, t) = -\frac{\partial}{\partial x}[f(x)p(x, t)] + \frac{1}{2}\frac{\partial}{\partial x}\left(g(x)\frac{\partial}{\partial x}[g(x)p(x, t)] \right). \tag{5.47}$$

In these expressions, we added brackets and parenthesis in order to make clear on what the operator act. For example, the term $\frac{\partial}{\partial x}[f(x)p(x, t)]$ has to be understood as

$$\frac{\partial}{\partial x}[f(x)p(x, t)] = \frac{\partial f(x)}{\partial x}p(x, t) + f(x)\frac{\partial p(x, t)}{\partial x}. \tag{5.48}$$

6
Stochastic models of growth

The demographic and the economic growth of a city is an extremely complex, multi-parametric and difficult to predict process. Historical examples of cities whose destinies have changed in very short time scales—one can think of San Francisco during the Gold Rush—show that they are objects almost as unpredictable as stock market prices (Batty, 2006). Therefore, the mathematical modeling of urban growth cannot reasonably be done by a deterministic equation; it calls for a statistical description, in the form of a *stochastic* process. For that reason, models describing the dynamics of city population are evaluated within a *set* or *system of cities* whose statistical properties are studied.

Models of city growth have historically emerged alongside the discovery of Zipf's law (Auerbach, 1913; Zipf, 1949), stating that in any country, the most populous city is generally twice as large as the next largest, and so on, a signature of the very high heterogeneity of city size (see Chapter 3). Zipf's law became the main paradigm of urban population statistics, and was even considered as a way to assess the validity of a urban model: models of city growth have to converge to Zipf's law law to be considered valid (Carroll, 1982). Only a few models in the 1970s (Haran and Vining Jr, 1973b; Vining Jr, 1974) have questioned the relevance of both Zipf's law for cities and the corresponding growth models. But apart from these last examples, most models of city growth propose mechanisms aiming at reproducing Zipf's law and therefore fall within the broad research field of power-law generation (Kumamoto and Kamihigashi, 2018).

In the following, we will mostly deal with models of multiplicative growth (or preferential attachment) and will not dwell on other models that have been proposed to explain urban growth like entropy maximization (Berry and Garrison, 1958) or allometric models (Beckmann, 1958). We will show that the most familiar multiplicative models are related and often similar, not to say identical (Moran, 2020). We will present several models which can pertain to city growth, even if they were originally developed in a various range of phenomena (number of species in a genus, number of words in a book, etc.).

In that sense, if this chapter is primarily a review of growth models for cities, it can also be seen as a review of multiplicative growth processes in general. We will not present the models in a chronological way but rather by pooling together models that are mathematically alike. A chronological description would, anyway, be of no avail, since models of multiplicative growth have been discovered and rediscovered several times in the last two centuries (Simkin and Roychowdhury, 2011).

The basic idea that justifies the use of multiplicative growth processes for cities is that city population grows, on average, exponentially.

In a discrete time representation, we then define a multiplicative process as a stochastic Markovian process where the population S_i of a given city i a time t evolves as

$$S_i(t+1) = \gamma_i(t)S_i(t), \tag{6.1}$$

where $\gamma_i(t)$ is a random variable whose mean $\mathbb{E}(\gamma_i(t))$ is assumed to be independent of $S_i(t)$.

In the following, we will sort multiplicative models of urban growth, all related by Eq. 6.1, into two different categories:

- Simon-based models (or "weak Gibrat" models). They remained the main paradigm of urban growth until the 1990s. They can be shown to obey Eq. 6.1 but with size-dependent variance. They are able to generate Zipf's law.
- Gibrat-based models (or "strong Gibrat" models). They became dominant in the 1990s in an adapted version. They assume that the variance of $\gamma_i(t)$ is independent of S_i. They do not directly generate Zipf's law unless a few supplementary assumptions are added.

Both categories have their limitations which have been pointed out in early works and confirmed more recently by empirical data. Nonetheless, an understanding of them remains paramount in urban studies.

The Yule–Simon model of growth

Yule's process

In 1925, George Udny Yule, a British statistician, proposed a stochastic process to explain the emergence and growth of species and biological genera (Yule, 1925). His model was based on previous work and data he had collected with the botanist John Christopher Willis in 1922 (Willis and Yule, 1922). Together, they proved that the distribution of biological genera followed a power-law.

Yule's paper was the first to derive a mathematical model for the emergence of Zipf's law based on random proportional growth. The original model considered N genera of a biological family where each genus initially contains one species. The dynamics are governed by various mutations. We adapt this model to urban growth and consider N cities initially containing one individual. The mapping between the original model and the city framework is then the following:

$$\text{family} \longleftrightarrow \text{system of cities}$$
$$\text{genus} \longleftrightarrow \text{city}$$
$$\text{species} \longleftrightarrow \text{individual}$$

At each step k of time dt, any individual in a city can generate a new individual in the same city with probability p (where $p^2 \ll p$). Introducing the continuous time variable $t = kdt$, the probability of a given city growing from n to $n+1$ individuals at time t is then

$$\mathbb{P}(n \rightarrow n+1) = pn = sndt, \tag{6.2}$$

where s is the growth rate so that $p = sdt$. This justifies the notion of preferential attachment: the probability of growth is proportional to the size of the city.

It is easy to show that the average fraction of cities with exactly one individual after k time steps is

$$f_1(k) = (1 - p)^k. \tag{6.3}$$

Since we have $pk = st$, we obtain

$$f_1(t) = \left(1 - \frac{st}{k}\right)^k \sim e^{-st}. \tag{6.4}$$

We denote by $f_n(t)$ the proportion of cities of size n at time t, and on average we have the following relation:

$$f_n(t + dt) = \underbrace{(1 - p)^n f_n(t)}_{\text{no new individual}} + \underbrace{(n - 1)p(1 - p)^{n-2} f_{n-1}(t)}_{\text{exactly one new individual}} + o(p^2), \tag{6.5}$$

from which we can derive by induction that

$$f_n(t) = e^{-st} \left(1 - e^{-st}\right)^{n-1}. \tag{6.6}$$

The average number of individuals per initial cities is then

$$n(t) = e^{st}. \tag{6.7}$$

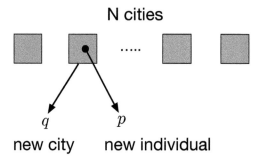

N cities

q p

new city   new individual

Fig. 6.1: Illustration of Yule's model (adapted to cities). An existing city gives rise to a new city with probability q, and an individual (shown by a black dot) in the city can give rise to a new individual with probability p.

Considering now that the number of cities can also grow with a probability q of each existing city generating a new city at each time step (see Fig. 6.1), one can show with the same argument (and by writing $gdt = q$) that the number of cities at time t is on average given by

$$N(t) = Ne^{gt}, \tag{6.8}$$

where g is proportional to the probability q generating a new city and such that

$$\mathbb{P}(N \to N + 1) = gN\mathrm{d}t. \tag{6.9}$$

Eq. 6.7 and Eq. 6.8 show that the number of cities grows as $\exp(gt)$, whereas the number of people in a given city grows on average as $\exp(st)$. Since the number of cities is now increasing, Eq. 6.6 is no longer the total fraction of cities of size n but the fraction of cities of size n of the *same age*—that is, cities that appeared at the same moment.

From Eq. 6.8 the expected number of new cities during time $\mathrm{d}t$ is $Ng\exp(gt)\mathrm{d}t$ and the fraction of cities of age x at time t is

$$m(x) = \frac{Nge^{g(t-x)}\mathrm{d}x}{Ne^{gt}} = ge^{-gx}\mathrm{d}x. \tag{6.10}$$

In order to compute the absolute distribution of sizes f_n, one needs to integrate over all cities of all possible ages. The absolute fraction of cities of size n is then given by

$$
\begin{aligned}
f_n &= \int_0^\infty \mathrm{d}t \; ge^{-gt} f_n(t) \\
&= g \int_0^\infty \mathrm{d}t \; e^{-(g+s)t} \left(1 - e^{-st}\right)^{n-1} \\
&= \frac{g}{s} \int_0^\infty \mathrm{d}x \; x^{\frac{g}{s}} \left(1 - e^x\right)^{n-1} \\
&= \frac{g}{s} \frac{\Gamma\left(\frac{g}{s} + 1\right)\Gamma(n)}{\Gamma\left(\frac{g}{s} + 1 + n\right)},
\end{aligned}
\tag{6.11}
$$

where Γ is the Gamma function. In the large n limit, we obtain

$$f_n \propto \frac{1}{\zeta}\Gamma\left(\frac{1}{\zeta} + 1\right) n^{-1-\zeta} \quad \text{with} \quad \zeta = \frac{g}{s}. \tag{6.12}$$

The fraction of cities of size n thus decreases as a power-law with exponent $1 + \zeta$.

Yule's model is thus the first historical mechanism of preferential attachment that is able to explain the emergence of Zipf's law (with parameter given here by ζ). In such a process, older cities constitute a larger fraction of the total number of individuals, and in that sense *aging* leads to a power-law distribution of city populations.

Simon's model

Simon's model (Simon, 1955) was developed to explain the emergence of Zipf's law in the distribution of word frequencies, an empirical fact first observed by Jean-Baptiste Estoup (1916) and later by George Zipf (1949).

Simon's demonstration applies to words and the number of their occurrences in books but can easily be translated to cities through a change of lexicon. A given word corresponds to a city and a specific occurrence of a word to an individual, and the dictionary is:

$$\text{word} \longleftrightarrow \text{city}$$
$$\text{number of occurrences of a word} \longleftrightarrow \text{population of a city}$$
$$\text{number of words in a book} \longleftrightarrow \text{time (or total population)}$$

The main mechanism here is the preferential attachment: over time, the probability that a new individual will move to a given city is proportional to its number of inhabitants. We consider a growing system of k individuals and we denote by $f(i, k)$ the number of cities with population i (when the total population is k, which can also be seen as the number of steps when one individual is added at each time step). Simon makes two assumptions:

1. The probability that the $(k + 1)$-th individual will move to a city which has a population i is proportional to $if(i, k)$, which is the *total* population of *all* cities that have population i after k steps. As explained by Simon, this assumption is weaker than the (more common) one stating that the probability that the $(k+1)$-th individual will move to a city which has a population i is proportional to i. Yet, the strong assumption implies the weak one.
2. There is a constant probability α that the $(k + 1)$-th individual will join a new city. As noted by Simon, one should think of α as the probability that a small settlement will reach a minimal population threshold that will make it a city.

Neglecting fluctuations for the steady state (i.e., assuming that $\mathbb{E}(f(i, k)) = f(i, k)$ where $\mathbb{E}(f(i, k))$ is the expectation of the number of cities with population i at time k), we can write the evolution equation under the form

$$f(i, k + 1) - f(i, k) = K(k)\left[(i - 1)f(i - 1, k) - if(i, k)\right], \tag{6.13}$$

where $K(k)$ is a proportionality factor. The first term of the right-hand side represents the case where the $(k+1)$-th individual joins a city which has population $i-1$ and the second term the case where the $(k+1)$-th individual joins a city which has population i. We have in addition the equation for the emergence of new cities (with population size equal to 1)

$$f(1, k + 1) - f(1, k) = \alpha - K(k)f(1, k) \tag{6.14}$$

with $0 < \alpha < 1$. Thus, we must solve the system

$$\begin{cases} f(i, k + 1) - f(i, k) = K(k)\left[(i - 1)f(i - 1, k) - if(i, k)\right] \\ f(1, k + 1) - f(1, k) = \alpha - K(k)f(1, k). \end{cases} \tag{6.15}$$

Since $K(k)if(i, k)$ is the probability that the $(k+1)$-th chosen city has population i, the value of the proportionality coefficient $K(k)$ is the solution of

$$\alpha + \sum_{i=1}^{k} K(k)if(i, k) = 1. \tag{6.16}$$

The total population is given by

$$k = \sum_{i=1}^{k} i f(i,k),$$ (6.17)

which implies

$$K(k) = \frac{1-\alpha}{k}.$$ (6.18)

By solving the system Eq. 6.15, one can show that the relative frequency of population size i (independent of k) is

$$f(i) = AB(i, \zeta + 1),$$ (6.19)

where B is the beta function and $\zeta = 1/(1-\alpha)$. In the limit $i \to \infty$, this leads to

$$f(i) \sim \frac{\Gamma(\zeta + 1)}{i^{1+\zeta}}.$$ (6.20)

This model thus predicts a distribution in agreement with Zipf's result with exponent $1/\zeta$.

Equivalence between Yule's and Simon's models

The equivalence between the two models was established by Simon himself in his original paper (Simon, 1955). To make it more obvious, we will use Simon's strong preferential attachment assumption, where the probability that the $(k+1)$-th individual will move to a *particular* city with population i is proportional to i. If we consider the size m of a given city, we have

$$\mathbb{P}(m \to m+1) = K(k)m.$$ (6.21)

Using Eq. 6.18, this gives

$$\mathbb{P}(m \to m+1) = (1-\alpha)\frac{m}{k} = (1-\alpha)\frac{m}{k}\mathrm{d}k$$ (6.22)

with $\mathrm{d}k = 1$. The important step to prove the equivalence between the two models is to introduce a "real" time variable t alongside the model time "k". Since k, the number of steps, is also the total population, we can write

$$k = e^{\gamma t},$$ (6.23)

where γ is the growth rate of the total population. This gives

$$\mathrm{d}k = \gamma k \mathrm{d}t$$ (6.24)

and we have

$$\mathbb{P}(m \to m+1) = (1-\alpha)m\gamma \mathrm{d}t.$$ (6.25)

We thus recover Eq. 6.2 from Yule's model by identifying

$$s = (1-\alpha)\gamma.$$ (6.26)

Besides, by definition, the number of different cities d obeys the growing equation

$$\mathbb{P}(d \to d+1) = \alpha dk. \tag{6.27}$$

On average, this leads to

$$d = \alpha k = \alpha e^{\gamma t}, \tag{6.28}$$

so that the growth rate of the number of different cities is $g = \gamma$ by equivalence with Eq. 6.8 from Yule's model. With Eq. 6.26, this last result proves that

$$s = (1-\alpha)g. \tag{6.29}$$

This leads to the equivalence between Simon's and Yule's models where Zipf's exponent is

$$\zeta = \frac{g}{s} = \frac{1}{1-\alpha}, \tag{6.30}$$

with s the growth rate of extant cities and g the growth rate of the number of cities. The two models are therefore similar up to a change in the time definition: if cities are added exponentially in Simon's process, it becomes equivalent to Yule's model. Interestingly, this same mechanism was found by Enrico Fermi to explain the empirical appearance of power-laws in the energy spectrum of cosmic radiations (Fermi, 1949).

We also note that both models rely on preferential attachment, since the probability of generating a new individual in a city is proportional to its population. Let $S_i(t)$ be the size of city i at time t, and we have

$$S_i(t+dt) = \begin{cases} S_i(t) + 1 & \text{with probability} \quad \alpha S_i(t)dt \\ S_i(t) & \text{with probability} \quad 1 - \alpha S_i(t)dt. \end{cases} \tag{6.31}$$

We define

$$\gamma_i(t) = \begin{cases} 1 + 1/S_i(t) & \text{with probability} \quad \alpha S_i(t) \\ 1 & \text{with probability} \quad 1 - \alpha S_i(t), \end{cases} \tag{6.32}$$

where we choose $dt = 1$ so that $S_i(t+1) = \gamma_i(t)S_i(t)$. This gives

$$\mathbb{E}(\gamma_i(t)) = 1 + \alpha \quad \text{and} \quad \text{Var}(\gamma_i(t)) = \alpha \left(\frac{\alpha}{S_i(t)} - 1 \right). \tag{6.33}$$

Simon's model is a multiplicative growth model that obeys Eq. 6.1 with average growth independent from $S_i(t)$ but with a variance that decreases with the population size. In that sense, it can be seen as "weak" version of Gibrat's law.

Connections with other processes

Interestingly, the Yule-Simon model is related to several other models of growth in physics and mathematics. These models are very similar, not to say identical, up to a change of words or variables (see Table 6.1).

Table 6.1: Equivalence between the lexicon of city growth and different models of multiplicative growth in different contexts. Genera in Yule's process, distinct words in Simon's model, nodes in Barabási–Albert's model, quantum states in Bose–Einstein statistics and families in Galton–Watson branching processes play the role of cities.

Model	Yule	Simon	Barabási–Albert	Bose–Einstein	Galton–Watson
city	genus	distinct word	node	quantum state	family
individual	species	word	link	particle	individual

Barabási–Albert model. In random networks, the Barabási–Albert model (Albert and Barabási, 2002) is a model for generating scale-free networks based on preferential attachment. Starting from m_0 initial nodes, a new node with $m < m_0$ links is added to the system at each time step. The probability $p(i)$ of the new node being connected to an existing node with degree k_i is then given by

$$p(i) = \frac{k_i}{\sum_j k_j}. \tag{6.34}$$

Hence, the probability of attaching to a specific node is proportional to its degree, implying that most connected nodes are preferred. It is a specific version of Simon's process where nodes play the role of distinct words (or cities) and the number of edges corresponds to individual words (or individuals). It yields a stationary probability distribution for the degree of nodes given by (Dorogovtsev and Mendes, 2002),

$$p(k) \propto \frac{1}{k^3}, \tag{6.35}$$

in the limit $k \gg 1$.

Bose–Einstein statistics. In quantum physics, Bose–Einstein statistics describes one of two possible ways in which a collection of non-interacting and indistinguishable particles may occupy a set of available discrete energy states. Particles that follow Bose–Einstein statistics are bosons and do not obey the Pauli exclusion principle: several particles can remain in the same quantum state. Consider L states with the same energy level ϵ. Bose–Einstein statistics is derived from the possible ways of arranging N particles in L states, particles being indistinguishable with each distinguishable arrangement being equiprobable.

In 1974, Bruce M. Hill justified city growth by using a Bose–Einstein-like process where people (like particles) are equiprobably distributed across cities (quantum states) in a distinguishable way (Hill, 1974). He showed that Zipf's law is a limiting distribution of Bose–Einstein statistics. Ijiri and Simon (1975) discussed Hill's result and proved in 1977 that if not only the number of particles but also the number of states grows with time, Bose–Einstein statistics is equivalent to Simon's model, a result rediscovered by Bianconi and Barabási over 20 years later (Bianconi and Barabási, 2001).

Branching processsses. A branching process is a stochastic process in a population with finite initial number of individuals and where each individual at generation t can breed new individuals at generation $t+1$ with some probability. The most famous branching process is the Galton–Watson process, discovered in 1870, although a previous description dates back to Irénée-Jules Bienaymé in 1845 and a solution to Antoine-Augustin Cournot in 1847 (Cournot, 1847).

In a Bienaymé–Galton–Watson process, each individual can give birth to n individuals between t and $t+1$ with probability $p(n)$ and then die. In this morose description of life, a family can become extinct at each generation with a non-zero probability, which is the reason why Galton, who was obsessed by eugenics and the extinction of aristocratic British families, developed the model.

The mathematical formulation of the Galton–Watson process is as follows. We denote by Z_n the number of individuals at generation n and by $X_{n,i}$, a random variable that denotes the number of direct successors of the individual i of generation n (the $X_{n,i}$ are generally assumed to be independent and identically distributed random variables). The general recurrence equation is then

$$Z_{n+1} = \sum_{i=1}^{Z_n} X_{n,i} \tag{6.36}$$

(and $Z_0 = 1$).

The extinction probability of the family is then $p_{\text{ext}} = \mathbb{P}(Z_n = 0)$ for large n. Either the single individual alive at time 0 has no offspring (which happens with probability $p(0)$), or each of their offspring generates a Galton–Watson process that reaches extinction. We can therefore write

$$p_{\text{ext}} = p(0) + \sum_{k=1}^{\infty} p(k)[p_{\text{ext}}]^k \tag{6.37}$$

and we recognize in the right-hand side of this equation the generating function of $X_{n,i}$ given by

$$f(z) = \sum_{n=0}^{\infty} p(n)z^n. \tag{6.38}$$

The extinction probability is then the fixed point of the generating function

$$p_{\text{ext}} = f(p_{\text{ext}}), \tag{6.39}$$

Defining the average number of children $\lambda = \sum_{n=0}^{\infty} p(n)n$, the branching process is called sub-critical with $p_{\text{ext}} = 1$ when $\lambda < 1$. When $\lambda > 1$, however, $p_{\text{ext}} < 1$, and while on average a share p_{ext} of families becomes extinct, the remaining families keep growing exponentially. Replacing individuals with species and families with genera, the Bienaymé–Galton–Watson process is similar to Yule's model, with the distinction that individuals are mortal while species are not.

A realistic distribution for the size of the offspring is a Poisson distribution of parameter λ:

$$p(n) = \frac{\lambda^n e^{-\lambda}}{n!}. \tag{6.40}$$

Considering the total size of the family (thus neglecting death) at time t given by

$$S(t) = \sum_{t=0}^{\infty} N(t), \tag{6.41}$$

one can prove from the Poisson distribution that the distribution of the family size is, in the large s limit,

$$p(s) \simeq \frac{e^{-s(\lambda-1-\log \lambda)}}{\lambda\sqrt{2\pi}} \frac{1}{s^{3/2}} \tag{6.42}$$

which converges to a power-law

$$p(s) \propto \frac{1}{s^{3/2}}, \tag{6.43}$$

when $\lambda \to 1$.

Limitations and variations

The Yule–Simon model remained the paradigm accounting for the existence of Zipf's law in city demographics until the 1990s. In 1996, the economist Paul Krugman acknowledged the capacity for Simon's model to justify the existence of power-laws in city size distributions in contrast with existing economic models, which predicted some sort of optimal size for cities and hence no hierarchical structure (Krugman, 1996). Krugman explained, however, that Simon's model was unable to reproduce the correct value of Zipf's exponent—allegedly 1—since, from Eq. 6.30, the exponent is:

$$\zeta = \frac{1}{1-\alpha}, \tag{6.44}$$

where α is the rate of appearance of new cities. The exponent $\zeta = 1$ would appear only for $\alpha = 0$ (no new cities), which would correspond to an infinite convergence time (Krugman, 1996).

Indeed, the notion of convergence in Simon's process is not straightforward. Contrary to frictional models, on which we will elaborate later, Simon's model has infinite growth: cities grow indefinitely, in particular the largest ones. Then, saying that Simon's process converges actually means that from a certain moment onward, the distribution of the *shares* of population of cities is stable (but not the population). This seems realistic: cities grow but the rank order between them is somehow stable. In such a case, assuming a power-law distribution at large times, we have:

$$\text{(Share of population of cities larger than } S) \propto S^{1-\zeta} \tag{6.45}$$

and

$$\text{(Total population of cities larger than } S) \propto S^{1-\zeta} S_{\text{total}}. \tag{6.46}$$

where S_{total} is the total population of the system. As long as $\alpha > 0$ (and $\zeta > 1$), the share of the largest cities becomes *relatively* smaller and smaller as cities get

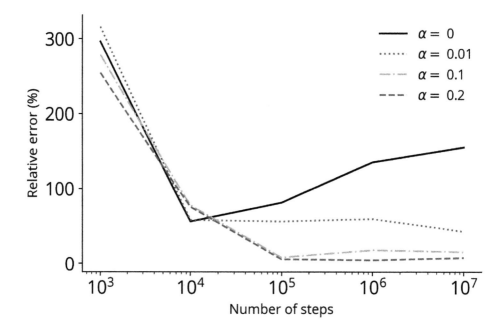

Fig. 6.2: Convergence time of Simon's process as a function of the probability of creating a new city α at each time step. As claimed by Krugman (1996), the convergence speed decreases when α goes to 0. Ultimately, Simon's model does not converge to a power-law when $\alpha = 0$, which makes it unable to generate Zipf's law with exponent 1.

larger. Their share eventually becomes negligible, justifying the steady-state assumption. When $\alpha = 0$, we have $\zeta = 1$, and the share of population of the largest cities does not decrease with size. Whatever S, the total population living in cities larger than S is constant. Ultimately, the same number of people live in cities larger than S and in cities larger than $2S$, which is absurd. By contradiction, one can prove that no steady state is reached. Simulations confirm Krugman's proof (see Fig. 6.2) and also that taking α to be small requires a long convergence period. In that sense, Krugman could refute Simon's model as being unable to yield Zipf's law with exponent 1 or close to 1 in a reasonable time frame.

This critique of Simon's model in the 1990s revived the interest in finding growth models yielding power-law distributions. Marsili and Zhang (1998) developed a model akin to Simon's model but with the possibility of the city population decreasing between two time steps. In this model, the number of cities $N(n, t)$ of size n at time t satisfies a master equation:

$$\partial_t N(n, t) = w_d(n + 1)N(n + 1, t) - w_d(n)N(n, t)$$
$$+ w_a(n - 1)N(n - 1, t) - w_a(n)N(n, t) + p\delta_{n,1}, \qquad (6.47)$$

where p is the probability of a new city emerging, $w_a(n)$ is the rate of population growth and $w_d(n)$ is the rate of population decrease of a city of size n. Taking $w_d(n) = 0$ would yield Simon's model.

From an interaction perspective, the increase and decrease rates are typically of a pair-wise type and $w \sim n^2$. The model is, hence, not exactly a multiplicative growth process, but under these conditions, it can be proven that

$$N(n) \propto \frac{p}{1-p} \frac{(1-p)^n}{n^2} \sim \frac{1}{n^2} \quad \text{for} \quad n \ll \frac{1}{p}, \tag{6.48}$$

which corresponds to Zipf's law with exponent equal to 1. The Zhang–Marsili assumption of n^2 interactions is, however, empirically incorrect (as we will show in Chapter 9, as the interaction term between cities is mostly governed by migrations that scale as n^β with $\beta < 1$).

Zhang and Marsili connected their model with another model of city formation stemming from physics: a reaction–diffusion model proposed by Zanette and Manrubia (1997). If we denote by $S_i(t)$ the population of city i at time t, the reaction process in this work is described by

$$S_i(t') = \begin{cases} (1-q)p^{-1}S_i(t) & \text{with probability } p \\ q(1-p)^{-1}S_i(t) & \text{with probability } 1-p, \end{cases} \tag{6.49}$$

with $t < t' < t+1$, $0 < p < 1$ and $0 \leq q \leq 1$. It is followed by a diffusion process (akin to migrations) described by

$$S_i(t+1) = (1-\alpha)S_i(t'). \tag{6.50}$$

This model proposed by Zanette and Manrubia appears, then, to be a multiplicative model with diffusion.

In addition to Krugman's comments on its convergence time, Gabaix discussed the main drawbacks of Simon's model acknowledging that it was the most plausible model for explaining city growth in the 1990s but that it still suffered from two main limitations (Gabaix, 2009):

1. From Eq. 6.30, a Zipf exponent equal to 1 implies $g = s$—that is, cities appear as fast as they grow, which is empirically incorrect (one expects $g < s$). This drawback is shared with the very simple Steindl's model (Steindl, 1965) in which cities grow at constant rate s and are generated at constant rate g without noise. Steindl's model generates a power-law with exponent g/s.

2. As we showed with Eq. 6.33, the variance of the growth rate decreases with the size, implying that larger units have a much smaller standard deviation of growth rate than small cities. Although the validity of Gibrat's assumption that variance is independent of size is not clear, Gabaix estimated that this was too strong a violation of empirical results to acceptance of Simon's model. It paved the way for the refinement of Gibrat's law in the 1990s (Gabaix, 1999), ending up in the formulation of Gabaix's model described below.

Both of these concerns will be proved to be incorrect. A value of 1 for Zipf's exponent is not a necessary feature of Zipf's law for cities, and the independence of

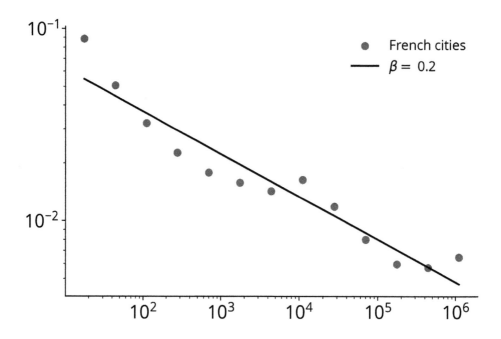

Fig. 6.3: Standard deviation $\sigma(S)$ of the growth rate of French cities between 2014 and 2015 as a function of the population S. The straight line corresponds to a power-law $\sigma(S) \propto S^{-\beta}$ with exponent $\beta \simeq 0.2 < 1$ ($R^2 = 0.91$), in contrast with Gibrat's law.

the variance of growth with respect to the size is an unclear result (Ioannides and Overman, 2003; Rozenfeld *et al.*, 2008), with contradictory estimates. For example, we plot on Fig. 6.3 the standard deviation of growth rate for French cities between 2014 and 2015 and find that

$$\sigma(S) \propto S^{-\beta} \tag{6.51}$$

with $\beta < 1$. This scaling law is similar to other results in economics, in particular for the growth of companies (Amaral *et al.*, 1997). Yet, the idea of switching from a weak version of Gibrat's law—in the form of Simon's process—to a strong (but modified) version of the law was critical in urban economics for the understanding of city growth. In the following, we will therefore elaborate on the application of Gibrat's law to cities.

Gibrat's law for cities

Gibrat's law

Robert Gibrat was a French engineer who took some interest in economics and proposed in 1931 the rule of proportionate growth (or law of proportionate effect), which states that the growth rate of a firm is independent of its size (Gibrat, 1931). More precisely, going back to Eq. 6.1, we will define Gibrat's law of proportionate effect, or the "strong" Gibrat's law, as the multiplicative process in which the population $S_i(t)$ of a city i evolves as

$$S_i(t+1) = \gamma_i(t)S_i(t),\qquad(6.52)$$

where $\gamma_i(t)$ is a noise independent of S_i (all quantities here are positive).

Defining the log-size $s_i(t) = \log S_i(t)$ as the relevant parameter of study, the multiplicative dynamics become additive:

$$s(t+1) = \log \gamma_i(t) + s(t)\qquad(6.53)$$

$$= \sum_{k=0}^{t} \log \gamma_i(t) + s_0.\qquad(6.54)$$

Assuming that the random variables $\gamma_i(t)$ are independent and identically distributed and behave well enough with finite mean $m = \mathbb{E}(\log \gamma_i(t))$ and variance $\sigma^2 = \mathrm{Var}(\log \gamma_i(t))$, one can use the central limit theorem to prove that $s(t)$ is distributed at large t according to a normal law

$$s(t) \sim \mathcal{N}(mt, \sigma^2 t),\qquad(6.55)$$

which implies that the population $S_i(t)$ follows a log-normal law with distribution

$$p(S,t) = \frac{1}{S\sqrt{2\pi\sigma^2 t}} \exp\left(\frac{-(\log S - mt)^2}{2\sigma^2 t}\right).\qquad(6.56)$$

This result raises several points of concern. Gibrat's law does not lead to a power-law and also doesn't reach a stationary distribution. The variance and expectation values of $S_i(t)$ diverge to ∞, whereas, as pointed out by (Moran, 2020), the most likely value behaves as $\exp((m - \sigma^2)t)$ and can go to 0.

However, one can convinced oneself that the log-normal distribution is *not too far* from a power-law (Sornette and Cont, 1997). More precisely, Eq. 6.56 can be rewritten as

$$p(S,t) = \frac{1}{\sqrt{2\pi\sigma^2 t}} S^{-1-\mu(S)} e^{\mu(S)mt}\qquad(6.57)$$

with

$$\mu(S) = \frac{1}{2\sigma^2 t}(\log S - mt).\qquad(6.58)$$

$\mu(S)$ is a slowly varying function of S which shows that the log-normal distribution can be mistaken for an apparent power-law distribution with exponent μ. Hence, Gibrat's law would not be far from yielding a power-law distribution if μ was almost constant, which would imply that $\log S$ was almost constant and S was not too big or too small. Hence, to ensure stationarity and convergence to a power-law behavior, one must add a way of preventing cities from becoming too large or too small: this is the main argument for introducing *friction* as an ingredient for preventing the emergence of cities that are too large or too small.

Friction

It is then natural to ask what (minimal) changes would need to be brought to Gibrat's law in order to reach a steady state with power-law distribution. Studying income dynamics, the English economist and mathematician (friend and colleague of Alan Turing) David Gawen Champernowne proposed a model of income growth yielding a power-law (Champernowne, 1953). We consider ranges of income R_0, R_1, R_2 so that $R_i = [y_i, y_{i+1}[$ with $y_{i+1} = hy_i$ (ranges of income are then constant on a log-scale). Consider a model in which, at each time step, there is a transition rate from one level of revenues to the other so that:

$$\mathbb{P}\left(R_i(t) \to R_j(t+1)\right) = \begin{cases} p_1 & \text{if } j - i = 1 \\ p_{-n} & \text{if } j - i = -n \\ 1 - p_1 - \sum_{k=1}^{n_{\max}} p_{-k} & \text{if } j = i \\ 0 & \text{otherwise,} \end{cases} \tag{6.59}$$

where $n_{\max}$ is the maximal number of moves allowed at a time. In this model, there is a probability of moving upward (in the income ladder) one level at a time and a probability of moving downward several steps a time. Champernowne added the assumption that

$$\sum_{k=n_{\max}}^{1} k p_k < 0, \tag{6.60}$$

which means that the average number of shifts per time unit is negative. This is the central stability assumption that prevents income from growing indefinitely and ensures the existence of an equilibrium. We also note that in this model, there is no way to go below the minimal income y_0, which acts as a sort of reflective barrier. Champernowne proved that a power-law distribution was a solution to this system.

As we will prove later, the two assumptions—the existence of a barrier and a negative drift—are paramount for ensuring a power-law. To understand that, we can go back to statistical physics. As Benoît Mandelbrot phrased it, the relevant variable of the multiplicative growth process obeying Eq. 6.1 is the log-size $w(t) = \log S(t)$, so that in the framework of $w(t)$, the dynamics is just a random walk or a diffusive process (Mandelbrot, 1960).

Imagine now a gas of particles in a vertical tube with no gravity, no top and no bottom. One expects the particles to diffuse to infinity without reaching a steady state. If there is a gravity field that acts as a negative drift, particles will escape to $-\infty$ in the absence of a bottom. Yet, if one adds a floor, particles will reach a steady-state distribution similar to the barometric distribution in the atmosphere, where the particles' height subject to gravity is distributed according to an exponential law (or Boltzmann law),

$$p(h) \propto e^{\frac{-E_{\text{pop}}(h)}{k_b T}} = e^{-\frac{mgh}{k_b T}}, \tag{6.61}$$

which yields a power-law in the original variable e^h. Hence, by preventing particles from escaping in both infinity directions, one can adapt a diffusive process to a converging one, reaching a power-law distribution. While a sketch of the proof was given by Levy and Solomon (1996) for a general noise, the case where the noise is Gaussian distributed can be easily solved in continuous time.

Geometric Brownian motion

In continuous time, Gibrat's law becomes a stochastic differential equation (SDE) with multiplicative noise, a geometric Brownian motion when the noise is Gaussian:

$$dS(t) = mS(t)dt + \sigma S(t)dB(t), \tag{6.62}$$

where $dB(t)$ is a Brownian motion. It can be written as a Langevin equation:

$$\partial_t S(t) = mS(t) + \eta(t)S(t), \tag{6.63}$$

where $\eta(t)$ is a Gaussian white noise with $\langle \eta(t) \rangle = 0$ et $\langle \eta(t)\eta(t') \rangle = \sigma_0^2 \delta(t - t')$. In Itô's convention, the probability distribution obeys the Fokker–Planck equation

$$\partial_t p(S,t) = -m\partial_S(Sp(S,t)) + \frac{\sigma_0^2}{2}\partial_S^2(S^2 p(S,t)). \tag{6.64}$$

Without further assumptions, the solution at time t is a log-normal distribution, as expected. If we now add friction to the system in the form of:

1. a negative drift $m < 0$,
2. a reflexive barrier in $S_{\min}$ so that Eq. 6.63 becomes

$$S(t + dt) = \max\left(S(t) + mS(t)dt + \sigma S(t)dB(t), S_{\min}\right), \tag{6.65}$$

the existence of the stationary solution is non-trivial (see (Harrison and Williams, 1987) for a demonstration). The stationary distribution obeys the steady-state Fokker–Planck equation

$$m\partial_S(Sp(S,t)) = \frac{\sigma_0^2}{2}\partial_S^2(S^2 p(S,t)), \tag{6.66}$$

whose solution is

$$p(S) = CS^{-1-\mu} \quad \text{and} \quad \mu = 1 - \frac{2m}{\sigma_0^2} \tag{6.67}$$

with C a constant. Since $m < 0$, we have $\mu > 1$. Defining the average city size

$$\bar{S} = \int_{S_{\min}}^{\infty} dS \; Sp(S) = \frac{\mu}{\mu - 1} S_{\min}, \tag{6.68}$$

we find

$$\mu = \frac{1}{1 - S_{\min}/\bar{S}}. \tag{6.69}$$

The exponent never reaches 1 but goes to 1 when the size barrier is very small.

Gabaix's model

Definition

At the end of the 1990s, Xavier Gabaix reviewed the different processes that were developed to explain Zipf's law for cities (Gabaix, 1999). He highlighted the afore-mentioned limitations of Simon's model:

1. Simon's model assumes that the number of cities grows as fast as the cities them-selves.
2. Simon's model assumes that the variance of growth decreases with city size.
3. Simon's model cannot account for an exponent value of 1 (convergence toward 1 would take infinite time (Krugman, 1996)).

Synthesizing Champernowne's early work and Levy and Solomon's reflective barrier with the theory of geometric Brownian motion, Gabaix proposed what became the new model of city growth,

$$S(t + dt) = \max\left(S(t) + dS(t), S_{\min}\right) \tag{6.70}$$

where $dS(t) = mS(t)dt + \sigma S(t)dB(t)$. This leads to a power-law,

$$p(S) = CS^{-1-\mu} \quad \text{and} \quad \mu = \frac{1}{1 - S_{\min}/\bar{S}}. \tag{6.71}$$

Taking infinitesimal friction ($S_{\min} \to 0$) yields Zipf's law with exponent 1.

To explain observed variations around the value 1 of the exponent, Gabaix proposed to study the effect of the dependence of variance growth on city size, giving the modified model

$$dS_t = \max\left(m(S)dt + \sigma(S)S_t dB_t, S_{\min}\right), \tag{6.72}$$

where $m(S)$ and $\sigma(S)$ depend on S. Under these conditions, the Fokker–Planck equation is

$$\partial_t p(S, t) = -\partial_S \left(m(S)p(S, t)\right) + \frac{1}{2}\partial_S^2 \left(\sigma_0^2(S)p(S, t)\right). \tag{6.73}$$

The steady-state solution obeys

$$\partial_S \left(m(S)p(S, t)\right) = \frac{1}{2}\partial_S^2 \left(\sigma_0^2(S)p(S, t)\right) \tag{6.74}$$

and has the local Zipf's exponent

$$\mu(S) = 1 - 2\frac{m(S)}{\sigma^2(S)} + \frac{S}{\sigma^2(S)}\frac{\partial \sigma^2(S)}{\partial S}. \tag{6.75}$$

Hence, the discrepancies around Zipf's exponent of value 1 can be explained by fluc-tuations around Gibrat's law.

Limitations

Gabaix's model is a very elegant way to understand urban growth while accounting for the existence of Zipf's law in population distributions. In contrast with Simon's model, it allows city populations to increase or decrease at any time.

One can, however, argue that there are several drawbacks in Gabaix's model. The most intuitive one is that one cannot perceive exactly what the meaning of a reflective barrier for cities is. Basically, this would mean that cities below a certain size cannot disappear and can bounce, in a sort of second-chance process. This implies that the number of cities remains constant and that the dynamics is conservative. This seems empirically false, as many cities have disappeared, and other models have been developed to account for the birth or death of a city. Another criticism that we will elaborate on later is that Gabaix's model is unable to account for the turbulent dynamics of cities through history (see Chapter 4 and (Batty, 2006)).

But more importantly, Gabaix's model, like previous models for urban growth, was developed with the purpose of generating Zipf's law. In that sense, it is not an empirical model of growth but a mathematical model that aims at converging to a power-law distribution of city population. However, as seen in Chapter 3, modern observations suggest that Zip's law does not hold in general. This gives far less credit to models whose only relevance was the convergence to Zipf's law. This limitation applies to the simplest models of growth, but even more strongly to complex models like Gabaix's equation. From an Occam's razor perspective, one cannot perceive the interest of introducing complexity, in the form of artificial friction, if converging to Zipf's law cannot be used for retrieving empirical data.

The discrepancy between empirical data and theoretical models of growth should lead us to think of new ways to look at population dynamics in cities. Before building our own model in Chapter 9, we will show in the next chapter that previous authors have tried to escape the paradigm of Zipf's law by resorting to a urban feature that cannot be overlooked: migrations. Migrations will be crucial in the remaining part of this book as the main criterion for understanding urban dynamics.

7
Models with migration

Interestingly, the models of urban growth seen so far do not explain where the growth comes from. They focus on explaining how cities grow, but not why. Without loss of generality, there are only two possible sources of population: natural growth (births minus deaths) and migrations, including migrations *between* cities. These migrations represent certain interactions and exchanges between cities and have been addressed in a small number of papers that we will discuss here.

A modified Yule–Simon model

In visionary, yet unfairly forsaken 1973 papers, Haran and Vining (Haran and Vining Jr, 1973a) highlighted that the Yule–Simon's model was developed in a context where the rural population was still very large, with people migrating from rural areas to cities, and where births exceeded deaths, implying that interurban migrations were not relevant (Haran and Vining Jr, 1973b). They developed the idea that such reasoning was false when births almost equal deaths or when migrations are mostly interurban; they considered this to be true in the 1970s, and this belief is still held today. In their model, they considered that "the dominant activity is intercity migration rather than migration from outside the system of cities to that system" which at that time was a rather visionary perspective about the statistics and dynamics of cities. Starting from Simon's process, they added the assumption that if the $t + 1$-th arrival is a migration, the probability that the individual will leave a city of size s is proportional to $sP(s,t)$ and the probability they will migrate to a city of size s' is proportional to $s'P(s',t)$ (where $P(s,t)$ is the probability that a city will have size s at time t). The proportionality factor in both cases is denoted by $K(t)$. The evolution of the probability distribution is then given by

$$P(s,t+1) = P(s,t) + (1-\beta)K(t)[(s-1)P(s-1,t) - sP(s,t)]$$
$$+ \beta K(t)[(s-1)P(s-1,t) - 2sP(s,t) + (s+1)P(s+1,t)], \qquad (7.1)$$

where β is the probability to have an interurban migration event. The initial condition (which corresponds to the creation of a new city whose probability is denoted by α) is given by

$$P(1,t+1) = P(1,t) + \alpha(1-\beta) - (1-\beta)K(t)P(1,t)$$
$$+ \beta K(t)[2P(2,t) - 2P(1,t)]. \qquad (7.2)$$

For $\beta = 0$, we recover Simon's process discussed above (see Chapter 6), which predicts a stationary distribution of the form

$$P(s, t \to \infty) \sim \frac{1}{s^\zeta}, \tag{7.3}$$

where $\zeta = 1/(1 - \alpha)$.

If we have migration effects characterized by a non-zero value of β, the steady-state distribution is no longer a pure power-law, as shown numerically in (Haran and Vining Jr, 1973a). Haran and Vining Jr indeed observed a curvature (in log–log) and a deviation from Zipf's law that increased with migration effects. In addition, they showed that this effect seemed to be present in US data (Haran and Vining Jr, 1973a).

In this respect, this paper is seminal both for being the first to question the relevance of Zipf's law in real data, and for invoking interurban migrations as paramount in explaining urban growth and the hierarchy of cities in a given country.

A master equation approach

Haag *et al.* (1992) who were focused mostly on the idea of modeling interurban migrations, proposed a stochastic framework and applied it to the system of French cities. They considered L cities and a hinterland h, with populations denoted by S_i, and the migration rate from city j to city i denoted by W_{ij}. The population configuration is described by the vector $\vec{s} = \{s_1, s_2, ..., s_L\}$, and the goal of this study was to describe the dynamics of this vector. The probability of an individual moving from i to city j is assumed to be of the form

$$p_{ij}(\vec{s}) = \nu_{ij}(t) e^{U_i(\vec{s}) - U_j(\vec{s})}, \tag{7.4}$$

where $\nu_{ij} = \nu_{ji}$ is the symmetric mobility factor and where U_i is the utility of city i and represents some sort of attraction potential.

The transition rate from city j to city i can then be written as

$$W_{ij}(\vec{s}) = s_j p_{ij}(\vec{s}) = s_j \nu_{ij}(t) e^{U_i(s_i+1) - U_j(s_j)}. \tag{7.5}$$

The complete master equation for the probability $P(\vec{s}, t)$ is, as often, too complicated, and the authors proposed to study the evolution equation for the average population of city k

$$\bar{s}_k(t) = \sum_{\vec{s}} s_k P(\vec{s}, t), \tag{7.6}$$

which reads as

$$\frac{d\bar{s}_k(t)}{dt} = \sum_{i=1}^{L} \bar{s}_i \nu_{ki} e^{U_k - U_i} - \sum_{i=1}^{L} \bar{s}_k \nu_{ik} e^{U_i - U_k}$$
$$+ W_{kh} - W_{hk} + \delta_k \bar{s}_k, \tag{7.7}$$

where W_{kh} and W_{hk} are transitions from and to the hinterland, δ_k the net growth rate of city k (births minus deaths), and the utilities U are computed over the average population $\bar{n}$.

These various parameters were estimated for France in the period 1954–1982, and this model was studied for the data in various scenarios to construct projections. In all cases, small rank variations were observed (with an almost constant top 10), and although it was not explicitly discussed in this paper, the distribution did not follow a simple Zipf's law (i.e. for the top 10, there is no simple relation between the rank and the population). This result confirms the earlier papers discussed above, with the inclusion of migration that is inconsistent with a simple power-law rank plot.

Diffusion with noise: The Bouchaud–Mézard model

The "regularization" of the Gibrat model proposed by Gabaix relies on the assumption that cities will be of a certain minimum size. However, we know the case of shrinking cities and that actually cities can even disappear. A different, simple way to "regularize" this behavior is to introduce migration effects between cities, and we will follow the presentation given by (Bouchaud and Mézard, 2000) in the study of wealth and adapt it to the case of cities. The equation for the population $S_i(t)$ of city i at time t is given by

$$\frac{dS_i}{dt} = \eta_i(t)S_i(t) + \sum_j J_{ji} - J_{ij}, \tag{7.8}$$

where the first term represents the "internal" growth and the last two terms represent movements between cities (see Fig. 7.1). The intercity flow can be rewritten as $J_{ij} = I_{ij}S_i$ and the equation becomes

$$\frac{dS_i}{dt} = \eta_i(t)S_i(t) + \sum_j \left(I_{ji}S_j(t) - I_{ij}S_i(t) \right), \tag{7.9}$$

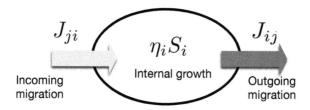

Fig. 7.1: Schematic representation of Eq. (7.8). There is an internal growth and "migrations" going either out or in, described by the matrix J_{ij}.

The random variables $\eta_i(t)$ are assumed to be identically independent Gaussian variables with the same mean m and variance given by $2\sigma^2$. The flow (per unit time) between cities i and j is $J_{ij} = I_{ij}S_j$, and for a general form of these couplings we are unable to solve this equation. We can, however, discuss the simple case of the complete graph where all units are exchanging with all others at the same rate taken equal to $I_{ij} = I/N$, where N is the number of cities (this scaling ensures a well-behaved large N limit). The equation for S_i becomes

$$\frac{dS_i}{dt} = I(\bar{S} - S_i) + \eta_i(t)S_i, \tag{7.10}$$

where $\overline{S}(t) = \sum_i S_i(t)/N$. We see that the first term acts as a homogenizing force towards the average $\overline{S}$. All cities feel the same environment which shows the mean-field nature of this simplified case. Formally, we can treat this equation as an equation for S_i subjected to a source $\overline{S}(t)$, and by integrating formally we obtain

$$S_i(t) = S_i(0)e^{\int_0^t (\eta_i(\tau)-I)d\tau} + I \int_0^t d\tau \overline{S}(\tau)e^{\int_\tau^t (\eta_i(\tau')-I)d\tau'}. \tag{7.11}$$

The average quantity $\overline{S}$ is still unknown at this point, and if we sum this equation over i, we still cannot solve it. However, if we assume that in the large N limit, the quantity $\overline{S}$ is self-averaging:

$$\overline{S} \simeq_{N\to\infty} \langle \overline{S} \rangle, \tag{7.12}$$

where the brackets $\langle \cdot \rangle$ denote the average over the variable η, we have

$$\overline{S}(t) \simeq \overline{S(0)}e^{(m+\sigma^2-I)t} + I \int_0^t d\tau \overline{S}(\tau)e^{(m+\sigma^2-I)(t-\tau)}, \tag{7.13}$$

where we assume that the initial conditions are independent from the η's. This equation can be easily solved (by, e.g., the Laplace transform) and we obtain

$$\overline{S}(t) = \overline{S(0)}e^{(m+\sigma^2)t}. \tag{7.14}$$

In order to observe a stationary distribution, we normalize S_i by $\overline{S}(t)$ and construct the variables $w_i = S_i(t)/\overline{S}(t)$. These quantities obey the Langevin-type equation

$$\frac{dw_i}{dt} = f(w_i) + g(w_i)\delta\eta_i(t), \tag{7.15}$$

where $f(w) = I(1-w) - \sigma^2 w$ and $g(w) = w$ (and $\delta\eta = \eta - m$). In order to write the Fokker–Planck equation for the distribution $\rho(w)$, we have to choose an integration rule for the multiplicative noise (see Chapter 5). Bouchaud and Mézard used the Stratonovich prescription (see Chapter 5) and obtained the Fokker–Planck equation for the distribution $\rho(w)$ under the form

$$\frac{\partial\rho}{\partial t} = -\frac{\partial}{\partial w}[f\rho] + \sigma^2\frac{\partial}{\partial w}\left[g\frac{\partial}{\partial w}g\rho\right]. \tag{7.16}$$

The equilibrium distribution for which $\partial\rho/\partial t = 0$ satisfies a simple equation, and once solved leads to

$$\rho(w) = \frac{1}{\mathcal{N}}\frac{e^{-\frac{\mu-1}{w}}}{w^{1+\mu}}, \tag{7.17}$$

where $\mathcal{N}$ is a normalization constant and where the exponent is given by

$$\mu = 1 + \frac{J}{\sigma^2}. \tag{7.18}$$

This result implies a number of consequences. First, when J is small (and non-zero), we observe that this regularization changes the log-normal distribution to a power-law (at large w) with an exponent between 1 and 2 (for J/σ^2 small). This distribution has a finite average $\langle w \rangle = 1$ but with an infinite variance, signalling large fluctuations.

We note here that an equation similar to Eq. 7.10 was previously proposed by Solomon (1998) in the context of generalized Lotka–Volterra equations, and subsequently developed in (Solomon and Richmond, 2001; Solomon and Richmond, 2002). In particular, Solomon and Richmond (2001) write a general equation of the form

$$\frac{dS_i}{dt} = [\eta_i + c_i(\{S_i\}, t)]S_i + a_i \sum_j b_j S_j \qquad (7.19)$$

where η_i is a Gaussian noise of zero mean and variance σ_i^2, and the c_i's are arbitrary functions. The last term of the right-hand side represents the redistribution between cities (in the original paper, the quantity of interest was wealth): the quantity $a_i b_j$ represents the fraction of the population in j that migrates to city i.

For arbitrary c_i factors, the dynamics of Eq. 7.19 lead to increasing inequalities. In particular, cities with very negative c_i eventually disappear. If all the c_i's are equal,

$$c_i(S_1, S_2, ..., S_N, t) = c(S_1, S_2, ..., S_N, t), \qquad (7.20)$$

Solomon and Richmond (2001) could show that the quantities $x_i = S_i/U(t)$ (where $U(t) = \sum_j b_j S_j$) are distributed according to the following distribution for constant σ_i and a_i ((Solomon and Richmond, 2001) and references therein):

$$P(x_i) \sim \frac{1}{x_i^{1+\alpha_i}} e^{-2a_i/x_i \sigma_i^2}, \qquad (7.21)$$

where the exponent is

$$\alpha_i = 1 + \frac{2a}{\sigma_i^2}, \qquad (7.22)$$

with $a = \sum_i b_i a_i$. This result shows that the mean-field result obtained by Bouchaud and Mézard is robust and does not necessitate strong assumptions about the different terms entering the general equation Eq. 7.19 to hold.

Applied to cities, this type of models provides a simple explanation for the diversity of Zipf's exponents observed in various countries (see (Soo, 2005)). The result is similar to what is obtained with Gabaix's model but here we obtain an exponent whose value depends on migrations between cities. Indeed, the diffusion model shows indeed that the origin of Zipf's law could lie in the interplay between internal random growth and exchanges between different cities. An important consequence is that increasing mobility should actually increase μ, and therefore reduces the heterogeneity of the city size distribution.

Few words of caution are, however, needed here: first, the relation Eq. (7.18) should be tested empirically, and second, the important assumption that I_{ij} is constant is by no means obvious, and we could imagine that this quantity actually depends on the distance $d(i, j)$, which could alter the results.

In the next part, we will see—in the spirit of the works mentioned in this chapter—that a more empirical account of migrations in city growth naturally leads to a new model of urban growth. This model, in the form of a stochastic differential equation of a new kind, highlights the role played by *rare events*, like one-off major migration waves, in urban dynamics.

How cities truly grow

8

The generalized central limit theorem and Lévy stable laws

In the previous part, through the review of models of urban growth, we saw how mathematical formalism can help with modeling city dynamics. These models were all stochastic: they do not predict the evolution of a specific city but give insights on the possible demographic outcomes of a system of multiple cities.

In this part, we will make up for the limitations of the models seen so far and we will propose a new approach to the mathematical description of city dynamics. This approach will rely on a more empirical and robust analysis of interurban migrations. But before developing our model, a few more mathematical results are necessary. In this chapter, we will start with a very common probabilistic result, the central limit theorem, and introduce its generalization. This will help us to define the so-called Lévy stable laws, which are typical "broad laws": laws that describe large but rare events. These laws obey their own rules and will be paramount in further descriptions of migrations flows. Indeed, we will see in Chapter 9 that interurban migrations flows are actually dominated by rare, but large, events.

The central limit theorem and its generalization

The central limit theorem is a cornerstone in the study of random variables. It proves that the addition of many random variables leads to a universal behavior, which is independent of the underlying random variables. Indeed, when independent identically distributed random variables add up, only their two first (if finite) moments, mean and variance, are relevant.

Besides its importance in probability and statistics, the central limit theorem is crucial in the study of social systems. Indeed, it shows that very few ingredients of individual behavior (described by a single random variable) play a role when summed up. In that sense, collective and universal structures, like cities or societies, can emerge from very different sets of individuals.

The existence of this theorem is therefore the foundation for understanding the statistical regularities observed in many complex systems. In this section, we will introduce this theorem and its demonstration, and we will also generalize it to random variables which display larger fluctuations.

The law of large numbers and the central limit theorem

Consider N identically independent distributed variables $\{X_i\}$ of finite average $m = \mathbb{E}(X_i)$ and finite variance $\sigma^2 = \mathbb{V}(X_i)$. The law of large numbers is a first step toward the understanding of the sum

$$S_N = \sum_i^N X_i. \tag{8.1}$$

and states that for every $\epsilon > 0$,

$$\mathbb{P}\left(|\frac{S_N}{N} - m| > \epsilon\right) \to 0. \tag{8.2}$$

This can be shown from Markov's inequality, valid for any non-negative random variable Y:

$$\mathbb{P}(Y > a) \leq \frac{\langle Y \rangle}{a}. \tag{8.3}$$

Applying Eq. 8.3 to the variable $Y = (S_N - mN)^2$ with $a = t^2$ leads to Chebyshev's inequality (Feller, 1957):

$$\mathbb{P}(|S_N - mN| > t) \leq \frac{N\sigma^2}{t^2}. \tag{8.4}$$

Taking $t = \epsilon N$ then leads to the law of large numbers.

A stronger probabilistic result is the central limit theorem. It states that

$$\mathbb{P}\left(\frac{S_N - Nm}{\sigma\sqrt{N}} < \beta\right) \to \mathcal{N}(\beta) = \frac{1}{\sqrt{2\pi}} \int_{-\infty}^{\beta} e^{-\frac{1}{2}y^2} dy, \tag{8.5}$$

which also implies that

$$P\left(\alpha < \frac{S_N - Nm}{\sigma\sqrt{N}} < \beta\right) \to \mathcal{N}(\beta) - \mathcal{N}(\alpha). \tag{8.6}$$

In other words, the distribution of the rescaled variable $[S_N - Nm]/\sigma\sqrt{N}$ converges in law toward a centered normalized Gaussian law:

$$\frac{S_N - Nm}{\sqrt{N}\sigma} \xrightarrow{\mathcal{L}} \mathcal{N}(0,1). \tag{8.7}$$

A simple demonstration relies on the use of the characteristic function

$$\varphi_X(t) = \mathbb{E}\left(e^{itX}\right). \tag{8.8}$$

The characteristic function of the sum of two random variables is simply the product of their characteristic functions. Hence, the characteristic function of the rescaled variable $U = (S_N - Nm)/\sigma\sqrt{N}$ is

$$\varphi_U(t) = \left[\varphi_X\left(\frac{t}{\sqrt{N}}\right)\right]^N. \tag{8.9}$$

When the two first moments of X exist, we can expand the characteristic function for small t:

$$\varphi_X(t/\sqrt{N}) = 1 - \frac{t^2}{2N} + o(1/N), \tag{8.10}$$

where $o(1/N)$ denotes a term going to 0 faster than $1/N$. We then obtain

$$\varphi_U(t) = \left[1 - \frac{t^2}{2N} + o(1/N)\right]^N \simeq e^{-t^2/2}, \tag{8.11}$$

with all the higher order terms vanishing for $N \to \infty$. We recognize here the characteristic function of the Gaussian distribution. The characteristic function being unique, this demonstrates the central limit theorem.

For this result to hold, two important assumptions should be made. First, the variables X_i are independent random variables (or at least the correlation function $\mathbb{E}(X_i X_j) - m^2$ should decrease fast enough with $|i - j|$). Second, the variables have a finite variance which allows the expansion of the characteristic function. In the case of broad distributions whose tails decay slowly (typically as power-laws with an exponent less than 2), the central limit theorem does not apply and other results have to be derived.

The most important point is the large N limit. Indeed, this theorem is valid for large enough N and the speed of convergence will depend on the distribution of the X_i. Distributions that are far from the Gaussian law (in particular with larger tails) will have a slower convergence speed. In fact, only the center of the distribution (hence the name "central limit theorem") of size of order $\sigma\sqrt{N}$ is well approximated by a gaussian function. In general, the size of the region well approximated by a Gaussian law depends on the distribution of the random variable considered. We refer the reader interested in this point to the discussion proposed in (Bouchaud and Potters, 2003).

Going beyond the central limit theorem

The discussion above shows that the central limit theorem does not apply for distributions with a diverging first or second moment. This is typically the case for power-law distribution behaving as

$$\rho(x) \sim \frac{1}{x^{1+\alpha}} \tag{8.12}$$

with $\alpha < 2$. The generalization of the central limit theorem to this case was proposed in (Gnedenko and Kolmogorov, 1954) elaborating on stable laws discussed by the French mathematician Paul Lévy (Lévy, 1926). Here, we will not follow these mathematical derivations but instead will present an elementary derivation proposed in (Amir, 2020).

This approach is based on a decimation idea (such as the one used in real-space renormalization) which states that the sum of random variables

$$S_N = \sum_{i=1}^{N} X_i \tag{8.13}$$

can be seen as a sum of n partial sums of m terms such that $N = m \times n$. This amounts to regrouping the X_i random variables in n packets of m variables each.

Our hope is that the distribution[1] of the normalized variable

$$\xi_N = \frac{S_N - b_N}{a_N} \tag{8.14}$$

converges to a well-defined limit when N is large:

$$\lim_{N \to \infty} \rho_{\xi_N}(x) = \rho(x). \tag{8.15}$$

Just like the roles played by $\sigma\sqrt{n}$ and m in the central limit theorem, the quantities a_n and b_n should determine respectively the width and the shift of the sought distribution.

We introduce the associated characteristic function of ξ_N and we assume that it is independent from N at large N:

$$\phi_{\xi_N}(\omega) = \phi(\omega) = \mathbb{E}\left(e^{i\omega x}\right). \tag{8.16}$$

After simple calculations, the characteristic function of S_N is

$$\phi_{S_N}(\omega) = e^{ib_N\omega}\phi(a_N\omega). \tag{8.17}$$

Using the identity $N = m \times n$ then allows us to write

$$\phi_{S_N}(\omega) = (\phi_{S_n})^m$$
$$= e^{imb_n\omega}\left(\phi(a_n\omega)\right)^m \tag{8.18}$$
$$= e^{i\frac{N}{n}b_n\omega}\left(\phi(a_n\omega)\right)^{N/n}. \tag{8.19}$$

For large N and n, this last expression should be independent from the choice of n, and assuming that n is large enough so that we treat this variable as a continuous one, we have:

$$\frac{\partial}{\partial n}e^{i\frac{N}{n}b_n\omega}\left(\phi(a_n\omega)\right)^{N/n} = 0. \tag{8.20}$$

Introducing $d_n = b_n/n$ and $\tilde{\omega} = a_n\omega$, the last expression can be shown to be:

$$\frac{\phi'(\tilde{\omega})\tilde{\omega}}{\phi(\tilde{\omega})} - \frac{a_n/n}{\partial a_n/\partial n}\log\phi(\tilde{\omega}) + i\tilde{\omega}n\frac{\partial d_n/\partial n}{\partial a_n/\partial n} = 0. \tag{8.21}$$

We introduce the following notations:

$$\begin{cases} \frac{a_n/n}{\partial a_n/\partial n} = C_1(n) \\ -n\frac{\partial d_n/\partial n}{\partial a_n/\partial n} = C_2(n). \end{cases} \tag{8.22}$$

The function ϕ is independent of n for large n. For Eq. 8.21 to hold for all n values, we have to ensure that the functions $C_1(n)$ and $C_2(n)$ have limits $\lim_{n \to \infty} C_1(n) = C_1$

---

[1] Here, we assume that this distribution of X_i exists.

and $\lim_{n\to\infty} C_2(n) = C_2$ respectively. If we assume that these constants are reached quickly, we can estimate the behavior for a_n and b_n. Indeed, writing

$$\frac{\partial a_n}{\partial n} = \frac{a_n}{n} \frac{1}{C_1} \tag{8.23}$$

$$\text{and} \quad \frac{\partial d_n}{\partial n} = -\frac{C_2}{n} \frac{\partial a_n}{\partial n} \tag{8.24}$$

leads to

$$\begin{cases} a_n & \propto n^{1/C_1} \\ b_n & \propto n^{1/C_1} + bn. \end{cases}$$

We note here that this solution is the leading order and corrections can be expected. Eq. 8.21 then reads

$$\frac{\phi'(\tilde{\omega})\tilde{\omega}}{\phi(\tilde{\omega})} - C_1 \frac{\log \phi(\tilde{\omega})}{\tilde{\omega}} = iC_2 \tag{8.25}$$

and introducing $u(\omega) = \log \phi(\omega)$ leads to

$$u'(\tilde{\omega}) - C_1 \frac{u(\tilde{\omega})}{\tilde{\omega}} = iC_2. \tag{8.26}$$

This differential equation is linear and its general solution can thus be written as the sum of the general solution of the homogeneous equation (with the right-hand side equal to 0) and a particular solution of the complete equation. The general solution of the homogeneous equation is

$$\begin{cases} u_{\text{hom}}(\tilde{\omega}) & = A_+\tilde{\omega}^{C_1} \text{ for } \tilde{\omega} > 0 \\ u_{\text{hom}}(\tilde{\omega}) & = A_-|\tilde{\omega}|^{C_1} \text{ for } \tilde{\omega} < 0, \end{cases} \tag{8.27}$$

where we note that A_+ and A_-, both complex numbers, are not necessarily equal.[2]
For a particular solution, we have to distinguish between the cases $C_1 \neq 1$, where the solution is of the form $u = D\tilde{\omega}$ with $D = iC_2/(1 - C_1)$ and the case $C_1 = 1$, where a particular solution is $iC_2\tilde{\omega} \log \tilde{\omega}$.
When $C_1 \neq 1$, the full solution in terms of ϕ is then given by

$$\phi(\omega) = e^{A|\omega|^{C_1} + iD\omega}. \tag{8.28}$$

The term C_1 plays the role of an exponent, and we will use the notation $C_1 = \alpha$. The term iD corresponds to a shift of the distribution. The quantity ϕ is the Fourier

---

[2]The differential equation is ill-defined when $\omega = 0$.

transform of a probability distribution which implies that $\phi(-\omega) = \phi^*(\omega)$ (where *
denotes the complex conjugation). We can thus rewrite ϕ as

$$\phi = \begin{cases} e^{A\omega^{C_1}} & \text{for } \omega > 0 \\ e^{A^*|\omega|^{C_1}} & \text{for } \omega < 0. \end{cases} \tag{8.29}$$

We finally obtain the solution that can be rewritten as

$$\phi(\omega) = \begin{cases} \exp\left(iD\omega - c|\omega|^\alpha[1 - i\beta\text{sign}(\omega)\tan(\pi\alpha/2)]\right) & \text{for } \alpha \neq 1 \\ \exp\left(iD\omega - c|\omega|^\alpha[1 + \frac{2i}{\pi}\beta\text{sign}(\omega)\log(\omega)]\right) & \text{for } \alpha = 1. \end{cases} \tag{8.30}$$

All possible attracting laws of the renormalized sum of random variables should be
(in Fourier space) of the same form as Eq. 8.30. They are called *Lévy stable laws*.

Lévy stable laws

Definition

In probability theory, a distribution is said to be stable if a linear combination of two
independent random variables following this distribution has the same distribution,
up to location and scale parameters. The stable distribution family is also sometimes
referred to as the Lévy α-stable distribution, after Paul Lévy, the first mathematician
to have studied it (Lévy, 1926).

The Lévy stable laws Eq. 8.30 depend on 4 parameters: the exponent α, the skewness β, the location parameter D, and the scale parameter c. We see on this form
Eq. 8.30 why this is the characteristic function of a stable law: the characteristic
function of the sum of two independent random variables is the product of the two
corresponding characteristic functions. Here, the product has the same values for β
and α but different values for the parameters D and c. The skewness β quantifies the
asymmetry of the distribution ($\beta = 0$ is a symmetric distribution).

The support of this function is $\mathbb{R}$, except when $\beta = 1$, in which case the support is
$[D, \infty[$. The mean is D (for $\alpha > 1$—otherwise it is infinite), and the median is given
by D when $\beta = 0$ (otherwise, it cannot be expressed in simple analytical terms).

We consider, as in (Amir, 2020), the case where $\rho(x) = A_+/x^{1+\alpha}$ for x large and
positive (with $\alpha < 1$) and $\rho(x) = A_-/|x|^{1+\alpha}$ for x large and negative—that is, when
the distribution has an asymptotic power-law behavior.

From the Tauberian theorems that connect the large x behavior of a function to
the small frequency behavior of its Fourier transform, the tails of the distribution
characterize the behavior of the characteristic function. The characteristic function
behaves for small ω as (using contour integration)

$$\phi(\omega) \simeq 1 - C\omega^\alpha, \tag{8.31}$$

where

$$C = \frac{\Gamma(1-\alpha)}{\alpha}(A_+ e^{-i\alpha\pi/2} + A_- e^{i\alpha\pi/2}). \tag{8.32}$$

It is then easy to show that

$$\frac{\text{Im}(C)}{\text{Re}(C)} = -\tan\left(\frac{\pi\alpha}{2}\right)\beta, \tag{8.33}$$

where β is defined as

$$\beta = \frac{A_+ - A_-}{A_+ + A_-}, \tag{8.34}$$

which illustrates this relation between the parameter β and the asymmetry of the distribution.

From Eq. 8.31, we also identify that the exponent α from the Lévy stable law is the same as the tail-exponent of the asymptotic power-law: Lévy α-stable laws behave asymptotically as power-laws of exponent α. Alternatively, a sum of (asymptotic) power-laws of exponent α can only converge to the Lévy stable law of parameter α.

Special cases

Unfortunately, there is no analytic expression in real-space of Lévy stable laws, except for some special cases that we discuss here.

Symmetric case. The simplest case is obtained for $\beta = 0$ which corresponds to the symmetric case around D. The characteristic function is then

$$\phi(\omega) = e^{iD\omega - c|\omega|^\alpha}, \tag{8.35}$$

which is a stretched exponential function. When $\alpha = 2$, we obtain the Gaussian distribution, and when $\alpha = 1$, we recover the Lorentzian (or "Cauchy") distribution

$$\mathbb{P}(x) = \frac{A}{x^2 + \pi^2 A^2}. \tag{8.36}$$

When α decreases from 2, the distribution becomes less peaked around its average and the tails become fatter. In other words, the random variable takes either small (or around average) values, or very large ones, and can therefore describe some sorts of intermittent processes (Bouchaud and Potters, 2003).

The normal distribution. The Gaussian distribution is a Lévy stable law with index $\alpha = 2$ and variance equal to $2c$, and mean D (β is irrelevant in this case). The characteristic function is

$$\phi(\omega) = e^{iD\omega - c\omega^2}. \tag{8.37}$$

This case corresponds to the standard central limit theorem. The sum of two independent random normal variables is normal, with mean the sum of means and variance the sum of variances.

The Lévy distribution.   The Lévy distribution is obtained for $\alpha = 1/2$ and $\beta = 1$ and is of the form (for $x \geq 0$)

$$\mathbb{P}(x) = \sqrt{\frac{C}{2\pi x^3}} e^{-C/2x}. \tag{8.38}$$

Its Fourier transform for $\omega > 0$ is given by

$$\phi(\omega) = e^{-\sqrt{-2iC\omega}}, \tag{8.39}$$

which corresponds to $\tan(1/2 \times \pi/2) = 1$, implying $\beta = 1$.

The Cauchy distribution.   The case $\alpha = 1$, $\beta = 0$ corresponds to the Cauchy distribution of the form

$$\mathbb{P} = \frac{1/\pi\gamma}{1 + \left(\frac{x-x_0}{\gamma}\right)^2}. \tag{8.40}$$

Its characteristic function is given by

$$\varphi(\omega) = e^{-\gamma|\omega|}. \tag{8.41}$$

This is clearly a stable function, as the characteristic function of a sum of such N variables is then

$$\varphi_N(\omega) = e^{-N\gamma|\omega|}. \tag{8.42}$$

The sum of such variables does not converge to a Gaussian but remains Lorentzian. In addition, we see that the width of the Lorentzian scales as N in contrast with the Gaussian case, where it scales as $\sqrt{N}$.

The generalized central limit theorem

We proved that the rescaled sum of random variables with infinite variance converges to a family of distributions whose Fourier transform is well defined and given by Eq. 8.30.

Since we saw in Eq. 8.25 that the rescaling parameters are $a_n \propto n^{1/\alpha}$ and $b_n \propto n^{1/\alpha} + bn$, we can now write the generalized version of the central limit theorem.

If $\{X_i\}$ are independent random variables identically distributed and having a power-law asymptotic behavior of exponent α, then

$$\sum_{i=1}^{n} X_i \xrightarrow{\mathcal{L}} \begin{cases} (bn)^{1/\alpha}\,\zeta_\alpha & \text{if} \quad \alpha < 1 \\ n\mu + (bn)^{1/\alpha}\,\zeta_\alpha & \text{if} \quad 1 < \alpha \leq 2 \\ n\mu + (bn)^{1/2}\,\eta & \text{if} \quad 2 < \alpha, \end{cases} \tag{8.43}$$

where ζ_α is a Lévy stable law of parameter α and η a Gaussian law.

In the generalized version, rare but large events dominate the behavior of the sum of the random variables. Indeed, it is possible that for a new random variable X_{N+1} we have the same order of magnitude as the sum

$$X_{N+1} \sim S_N, \tag{8.44}$$

in sharp contrast with finite variance variables where

$$X_{N+1} \ll S_N. \tag{8.45}$$

This highly theoretical result is at the heart of the new urban growth equation that is the subject of the book. In the next chapter, we introduce a bottom-up approach to urban growth in which intercity migration plays a major role. By showing that these migration shocks have an asymptotic power-law behavior of exponent $\alpha < 2$, the generalized central limit theorem will allow us to derive a new type of stochastic differential equation.

9

From first principles to the growth equation

So far, we have seen that most research about city growth has been conducted with the underlying idea that the stationary state of the urban population for a set of cities is described by Zipf's law (see Chapter 3).

In that respect, Zipf's law is considered to be a cornerstone of urban economics and geography, and maybe one of its few quantitative universal results. Previous mathematical attempts to model city growth have taken Zipf's law as a validation proof: the stationary distribution of city population in a system of cities had to follow Zipf's law to enforce the model consistency (see Chapter 6).

The consequences of this law and its explanatory models are multifold. Zipf's law states that the population distribution of urban areas in a given territory (or country) displays a Pareto law with exponent equal to 2 or, equivalently, that the city populations sorted in decreasing order versus their ranks follow a power-law with exponent 1. This result characterizes the hierarchical organization of cities, and implies that in any country, the city with the largest population is generally twice as large as the next largest, and so on. It is a sign of the very large heterogeneity of city sizes and shows that cities are not governed by optimal considerations that would lead to one unique size but, on the contrary, that city sizes are broadly distributed and follow some sort of hierarchy. Hence, Zipf's law states that there is no optimal city size, in contrast with economic models of city equilibrium—something Krugman called the "mystery of urban hierarchy". On the contrary, models that generate Zipf's law for cities fall into a broad category of mathematical models generating power-laws and lack economical ingredients: they are so universal that they explain how cities grow but not why.

More fundamentally, as Lord Kelvin framed it for physics a century ago, there are two "clouds" in Zipf's law theory and its associated growth models. The first one we have seen is that recent studies, supported by an increasing number of data sources, have questioned the existence of such a universal power-law and have shown that Zipf's exponent can vary around 1 depending on the country, the time period, the definition of cities used or, more crucially, the fitting method (see Chapter 3). The second one is that beyond understanding the stationary distribution of urban populations lies the problem of their temporal evolution. As already noted, the huge number of studies regarding population distribution contrasts with the few analyses of the evolution of cities over time. Cities and civilizations rise and fall many times on a large range of time scales, and Gabaix's model is both quantitatively and qualitatively unable to explain these specific chaotic dynamics (see Chapter 4). Therefore, a model able to simultaneously explain observations about the stationary population distribution and the temporal dynamics of systems of cities is missing. In particular, we are not at this point able to identify the causes of the diversity of empirical observations about the hierarchical organization of cities, the occurrence of megacities and the empirical instability in city dynamics seen in the births and deaths of large cities over short time scales. In this respect, we need not only a quantitative improvement in the models but a paradigm shift.

Building a bottom-up equation

To understand the growth of cities, we need a bottom-up approach, starting from elementary mechanisms and reproducing general universal trends. Without loss of generality, the growth dynamics of a city of size S_i belonging to a system (such as a country) can be decomposed into the sum of a natural growth term (births and deaths) and a migration term, the balance of the number of newcomers to the city minus the number of emigrants (see Chapter 7 and (Bettencourt and Zünd, 2020)). For convenience and to facilitate consistency with sources of migration data between cities, we propose to recombine these terms by separating migration within the $\{S_i\}$ system on the one hand, and natural growth and migration outside the system on the other:

$$\partial_t S_i = \underbrace{\text{Natural growth } + \text{ Migrations out of } \{S_i\}}_{\text{"out of system" growth}}$$

$$+ \underbrace{\text{Migrations within } \{S_i\}}_{\text{interurban migrations}}. \tag{9.1}$$

We can now study the "out-of-system" growth and interurban migrations separately.

Out-of-system growth

What we call out-of-system growth is the addition of two heterogeneous sources of growth: the natural growth of the city i and the migrations that do not take place within the system of cities, which correspond essentially to international migrations and exchanges between the city i and the hinterland (cities, towns or villages that are too small to be taken into account in $\{S_i\}$).

Setting aside these external sources is empirically not absurd, since they represent, at least nowadays, a minor contribution to city growth. In the US, between 2013 and 2017, more than 9 000 000 people moved from one metropolitan area to another, compared to 1 300 000 people abroad and 1 500 000 people in the rest of the US. Migrations within the city system are thus dominant.

Empirically, the "out-of-system" growth is well approximated by a normal law (see Fig. 9.1). In the continuous time limit, we will represent it as a stochastic growth process with multiplicative Gaussian Langevin noise $\eta(t)$.

It is interesting to note that in their visionary article questioning the relevance of Zipf's law for cities, Haran and Vining (see Chapter 7) noted that deviations from Zipf's law could be associated with the maturity of a system of cities, with the basis of

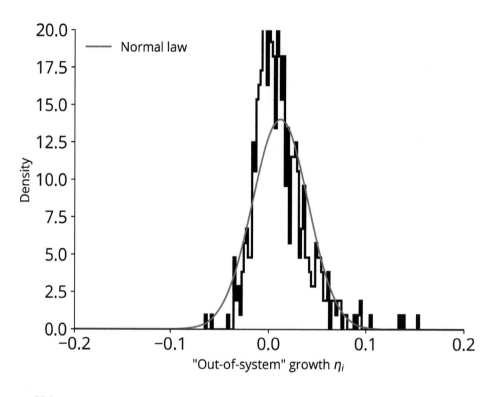

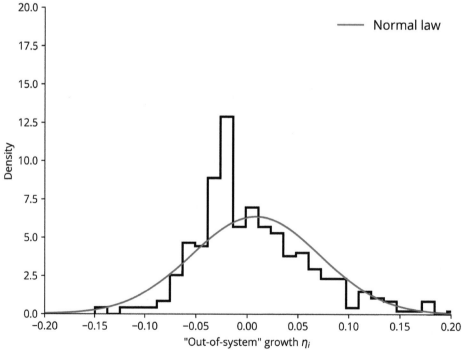

Fig. 9.1: The "out-of-system" growth of a system of cities is well approximated by a normal law. Example here from French (top) and US (bottom) cities.

maturity being that this "out-of-system" growth has become minor. Haran and Vining distinguished between two types of cities. First, developing cities where births exceed deaths and where the rural component is still large, leading to fast development. In this context, Simon's model of growth is quite relevant and the "out-of-system" growth is dominant, leading back to a Gibrat-like model of growth with only one contribution in Eq. 9.1. In the second type of cities, however, births and deaths are of the same order of magnitude and migrations from the hinterland low; in such cases, Simon's model is unfounded and we need a more complex framework, where interurban migrations have to be quantified.

Defining a graph of interurban flows

In the remainder of this chapter, we consider four recent migration datasets in four different countries (U.S. intercity migration between 2012 and 2017, French intercity migration between 2003 and 2008, UK intercity migration between 2012 and 2016, and Canadian intercity migration between 2012 and 2016). The data of migration flows between metropolitan areas provide a convenient framework for graph analysis. These interurban migrations can be represented as a weighted, oriented graph whose vertices are metropolitan areas. The link from i to j is then described by the flow $J_{i \to j}$. In this graph representation, two cities are called *neighbors* if at least one individual has moved from one to the other: $J_{i \to j} > 0$ or $J_{j \to i} > 0$. Since the graph is oriented, we must distinguish between *in*-neighbors and *out*-neighbors. In-neighbors of city i are cities from which new individuals have migrated toward i. Out-neighbors of city i are cities to which previous living-in-i individuals have moved.

We show in Fig. 9.2 the number $\mathcal{N}_{in}(i)$ of in-neighbors versus the number $\mathcal{N}_{out}(i)$ of out-neighbors. We have on average $\mathcal{N}_{out}(i) \sim \mathcal{N}_{in}(i)$ (see also Table 9.1). This result leads us to define an average number of neighbors for each city i in the following way:

$$\mathcal{N}(i) = \frac{\mathcal{N}_{out}(i) + \mathcal{N}_{in}(i)}{2}. \tag{9.2}$$

Table 9.1: Estimates of the mean and variance of the normalized quantity $(\mathcal{N}_{out}(i) - \mathcal{N}_{in}(i)) / \max(\mathcal{N}_{out}(i), \mathcal{N}_{in}(i))$ suggesting that we have approximately $\mathcal{N}_{out}(i) \sim \mathcal{N}_{in}(i)$.

Dataset	Mean	Variance
France	4%	5%
US	0.3%	5%

It is natural to assume that the number of neighbors $\mathcal{N}(i)$ depends on the population S_i, and we show the corresponding plot on Fig. 9.3. This plot suggests a power-law behavior of the form

$$\mathcal{N}(i) \propto S_i^{\gamma} \tag{9.3}$$

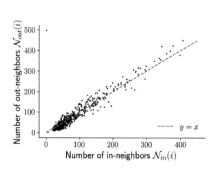
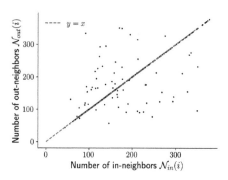

Fig. 9.2: Number of in-neighbors versus number of out-neighbors in France (left) and in the US (right). In-neighbors of city i are cities from which new individuals have migrated toward i. Out-neighbors of city i are cities to which individuals that previously lived in i have moved. We have on average $\mathcal{N}_{out}(i) \sim \mathcal{N}_{in}(i)$.

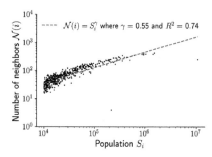
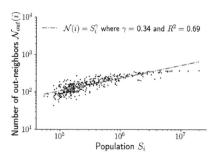

Fig. 9.3: The number of neighbors of a given city i scales a power-law of the population S_i of i with exponent $\gamma < 1$. In France (left), $\gamma = 0.55$ and in the US (right), $\gamma = 0.34$.

and the fit gives the following values for the exponent $\gamma \simeq 0.34$ and $\gamma \simeq 0.55$ for the US and France, respectively. In the UK and in Canada, the datasets are small (respectively 41 and 160 cities) and all cities are connected, which implies $\gamma = 0$.

We hence find a scaling law relating the number of neighbors of city i to its population: $|N(i)| \propto S_i^{\gamma}$, where $\gamma \simeq 1/2$. The temporal evolution of the population size S_i can then be written as

$$\frac{\partial S_i}{\partial t} = \eta_i S_i + \sum_{j \in N(i)} (J_{j \to i} - J_{i \to j}), \qquad (9.4)$$

where the quantity η_i is a random variable representing the "out-of-system" growth of city i, empirically distributed according to a normal distribution, and the flow $J_{i \to j}$ is the number of individuals moving from city i to city j during a time period dt.

This equation is a simple balance equation, very similar to the Eq. 7.8 of the Bouchaud–Mézard model (see Chapter 7). We also note that if there is an exact balance of migration flows ($J_{i \to j} = J_{j \to i}$), the equation becomes equivalent to the continuous version of the Gibrat model (Eq. 6.63), which predicts a log-normal distribution of populations.

Starting from the general Eq. 9.4 is very natural, since it amounts to summing up births, deaths and migrations in a rearranged way. Nevertheless, as is often the case when very general basic equations are used, it is difficult to use. Simplifications of this equation have been proposed, in which various assumptions lead to the Gibrat model, but lack the large fluctuations in the migration terms, which, as we shall see, are quantitatively and qualitatively crucial.

Gravitational model

A natural way to study flows appearing in Eq. 9.4 is to test the well-known gravitational model that in the case of migration from city i to city j reads as

$$J_{i \to j} = I_0 \frac{S_i^\mu S_j^\nu}{d_{ij}^\sigma}, \tag{9.5}$$

where d_{ij} is the geographical distance between i and j, μ, ν, σ are exponents usually determined by fitting the data and I_0 is a prefactor (also determined by the fit).

Since we aim at building a model of migrations that is as parsimonious as possible, we can test the relative importance of distances d_{ij} and populations S_i in explaining migrations.

To do this, we test two different models:

- the "full" gravitational model $J_{i \to j} = I_0 \frac{S_i^\mu S_j^\nu}{d_{ij}^\sigma}$
- a distance-free model $J_{i \to j} = I_0 S_i^\mu S_j^\nu$.

We fit the parameters of both models on French and US data using log-linear regressions and look at how much of the data dispersion is explained in the two cases. Results are shown on Table 9.2.

Table 9.2: Estimates of the gravitational model parameters obtained by log-linear regression with and without the effect of distance. Distances explain only a small part (about 8 %) of the dispersion of flows. We had no distance data for the UK and Canada.

Dataset	I_0	μ	ν	σ	R^2
US flows (distance included)	$1.6 \cdot 10^{-2}$	0.46	0.44	0.53	0.42
US flows (no distance)	$3.8 \cdot 10^{-3}$	0.38	0.37	/	0.34
France flows (distance included)	$3.7 \cdot 10^{-3}$	0.49	0.49	0.54	0.45
France flows (no distance)	$4.3 \cdot 10^{-4}$	0.45	0.45	/	0.38

We first observe that parameters for France and the US are very close ($\mu \in [0.38, 0.49]$, $\nu \in [0.37, 0.44]$ and $\sigma \simeq 0.5$), and that in the two countries both models explain about 40 percent of the dispersion (US: $R^2 = 0.42$ with distances and

$R^2 = 0.34$ without; France: $R^2 = 0.45$ with distances and $R^2 = 0.38$ without). The increase of the fit quality resulting from taking distances d_{ij} into account is small: the value of R^2 increases by 8 percent in the US case and 7 percent in the French one. Using a distance-free and population-dependent model thus gives results almost as good as a full gravitational model while being far more simple.

This analysis leads us to the conclusion that distance effects are not paramount in explaining migration flows. They can be discarded as being second-order corrections compared with city sizes which play the dominant role. This does not mean that we neglect distances but that they can be included in the noise. In this way, we can reduce our calculations and equations to a more robust and universal population-dependent model.

We also note that the gravitational model is overall very noisy, as most of the dispersion cannot be explained by populations or distances. The relevance of the gravitational model is a common subject of debate in the literature, and other models of human mobility seem to better explain empirical results. The gravitational model is, however, very simple to write and gives acceptable enough results for our analysis. Indeed, a key point of our reasoning is to note that, although fluctuations around the model $I_0 S_i^\mu S_j^\nu$ are very large, we can still deal with migration flows if we consider them as random variables. This is what we do while deriving our general equation.

Minimal model for the interurban migration flows

Average detailed balance

We thus write the average migration flow from i to j as $\langle J_{i \to j} \rangle = I_0 S_i^\mu S_j^\nu$, where I_0 is a constant prefactor. Introducing the rate per capita $I_{ji} = J_{i \to j}/S_i$, we can study the ratio of the two rates from i to j and j to i:

$$\frac{I_{ij}}{I_{ji}} = \left(\frac{S_i}{S_j} \right)^{1 - \mu + \nu}. \tag{9.6}$$

We plot this quantity versus S_i/S_j on Fig. 9.4 (left), which shows that $1 - \mu + \nu \simeq 1$, implying that $\mu \simeq \nu$. This result also implies that on average $\langle J_{i \to j} \rangle = \langle J_{j \to i} \rangle$, which is some sort of detailed balance condition: on average, flows and counterflows balance each other, a long-lasting result of human geography. Yet, this result is only true on average, and fluctuations are so big that typical flows between two cities would not compensate each other.

From these conclusions, we can build up a minimal model of migration flows:

$$\langle J_{i \to j} \rangle = I_0 S_i^\nu S_j^\nu \tag{9.7}$$
$$\text{and } \langle I_{ij} \rangle = I_0 S_i^\nu S_j^{\nu - 1}, \tag{9.8}$$

The fit of I_{ij} gives $\nu \simeq 0.4$ in the US and in France. Fluctuations are large and a log-likelihood model gives the following error bars: $\nu = 0.4 \pm 0.3$ for France and $\nu = 0.4 \pm 0.4$ for the US.

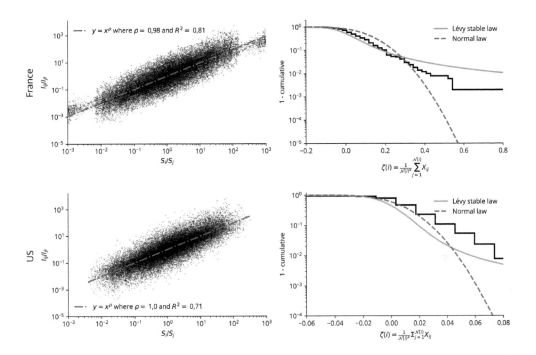

Fig. 9.4: An analysis of migration flows in France and the United States. On the left is the ratio of the migration rate to the population ratio. The line is a power-law regression that gives an exponent equal to 1. On the right is the empirical cumulative distribution function (right tail) of renormalized migration flows ζ_i compared with a Lévy distribution and the normal distribution.

Heavy-tailed fluctuations

This analysis leads us to write

$$I_{ij} = I_0 S_i^\nu S_j^{\nu-1} x_{ij} \tag{9.9}$$

where the random variable x_{ij} has average 1 and fluctuations that we assume to be large, an assumption confirmed *a posteriori* by the analysis presented in Fig. 9.4.

Thus, the term $J_{i\to j}$ can be represented by a variable of the form $I_0 S_i^\mu S_j^\nu x_{ij}$, where the random variables x_{ij} have mean equal to 1 and encode all the noise (which includes all sorts of hidden parameters including the distance). We denote by $I_{ji} = J_{i\to j}/S_i$ the probability per unit of time and per inhabitant of moving from city i to city j. The ratio I_{ij}/I_{ji} as a function of the population ratio S_i/S_j exhibits, on average, a linear behavior (Fig. 9.4, left column). This implies that $\mu = \nu$, and that we have, on average, a kind of detailed balance $\langle J_{i\to j}\rangle = \langle J_{j\to j}\rangle$ (the brackets $\langle ...\rangle$ refers to the average over the noise η), but that, crucially, the fluctuations are non-zero.

More precisely, if we define $X_{ij} = (J_{j \to i} - J_{i \to j})/I_0 S_i^\nu$, we observe that, empirically, the random variables X_{ij} have heavy tails—that is, they are distributed according to a broad distribution which decreases asymptotically as a power-law with an exponent $\alpha < 2$ (Fig. 9.5).

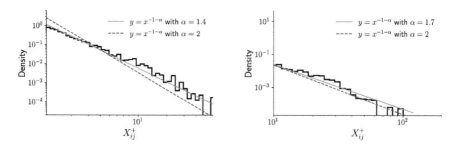

Fig. 9.5: Density of the right-tail quantity X_{ij}^+ (ie. $X_{ij} > 0$) in France (left) and in the US (right). Both distributions have heavy tails and the tail of both distributions is asymptotically described by a power-law with exponent $1 + \alpha$ where $\alpha < 2$.

Eq. 9.4 can thus be rewritten in the form

$$\sum_{j \in N(i)} (J_{j \to i} - J_{i \to j}) = I_0 S_i^\nu \sum_{j \in N(i)} X_{ij}. \tag{9.10}$$

Assuming that the correlations between the variables X_{ij} are negligible, the generalized central limit theorem (see Chapter 8) allows us to write that

$$\zeta_i = \frac{1}{|N(i)|^{\frac{1}{\alpha}}} \sum_{j \in N(i)} X_{ij} \tag{9.11}$$

follows a Lévy law L_α of parameter α for a sufficient large size of $N(i)$ (we note that Lévy laws have an asymmetry parameter, which is not detailed here but which is not empirically zero). This is confirmed empirically: French, US, UK, and Canadian data are better represented by a stable Lévy distribution than by any other distribution (Fig. 9.4, right column). The estimates of α are given in Table 9.3.

In particular, we recall that the important thing is to observe a value $\alpha < 2$. This leads then to

$$\frac{\partial S_i}{\partial t} = \eta_i S_i + D S_i^\nu N(i)^{\frac{1}{\alpha}} \zeta_i \tag{9.12}$$

$$= \eta_i S_i + D S_i^\nu S_i^{\frac{\gamma}{\alpha}} \zeta_i. \tag{9.13}$$

Thus, the basic Eq. 9.4 can be rewritten as a new kind of stochastic differential equation that is the focal point of this book:

$$\frac{\partial S_i}{\partial t} = \eta_i S_i + D S_i^\beta \zeta_i \tag{9.14}$$

Table 9.3: Estimation of the parameter α in four different ways in the four countries. A maximum likelihood estimate is compared with the parameter minimizing the Kolmogorov–Smirnov distance, the log-moment method and the Hill parameter.

Dataset	MLE	Kolmogorov–Smirnov	Log-moments	Hill
France 2003–2008	1.43 ± 0.07	$1.2 < \alpha < 1.8$	1.3	1.4 ± 0.3
US 2013–2017	1.76 ± 0.07	$1.7 < \alpha < 1.8$	1.6	Inconclusive
UK 2012–2016	1.32 ± 0.26	Inconclusive	1.0	1.2 ± 0.8
Canada 2012–2016	1.69 ± 0.12	Inconclusive	1.9	1.4 ± 0.6

where $D \propto I_0$ and $\beta = \nu + \gamma/\alpha$ and η_i is a Gaussian noise having the mean growth rate r and the dispersion σ. The value of β can be also be directly estimated from log-linear regression as made on Fig. 9.6.

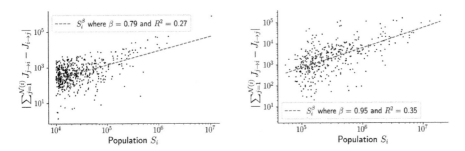

Fig. 9.6: Dependence of the total net migration of city i $\left| \sum_{j=1}^{\mathcal{N}(i)} J_{j \to i} - J_{i \to j} \right|$ on the population S_i. We fit the parameter β so that $\left| \sum_{j=1}^{\mathcal{N}(i)} J_{j \to i} - J_{i \to j} \right| \propto S_i^{\beta}$ in France (left) and in the US (right).

By studying how $\sum_{j=1}^{\mathcal{N}(i)} (J_{j \to i} - J_{i \to j}) = D S_i^{\beta} \zeta_i$ behaves as a function of S_i, we find $\beta = 0.93$, $\beta = 0.79$, $\beta = 0.51$ and $\beta = 0.78$ for the US, France, the UK and Canada respectively. We compare these values with the prediction of our model $\beta = \nu + \gamma/\alpha$ in Table 9.4.

Table 9.4: There is good agreement between measured and predicted β values in the US and France. The UK and Canadian datasets are small and fully connected (implying $\gamma = 0$) and therefore more noisy.

Dataset	γ	ν	$\beta = \nu + \gamma/\alpha$	β measured
France 2003–2008	0.55 ± 0.06	0.4 ± 0.3	0.8 ± 0.4	0.79 ± 0.07
US 2013–2017	0.34 ± 0.05	0.4 ± 0.4	0.6 ± 0.5	0.96 ± 0.05
UK 2012–2016	0	0.7 ± 0.3	0.7 ± 0.3	0.51 ± 0.05
Canada 2012–2016	0	0.5 ± 0.4	0.5 ± 0.4	0.78 ± 0.06

Correlation of noises

In Eq. 9.14, neither the noises nor the cities are correlated—that is, the increase in the number of inhabitants due to migration in one place does not mean that there is a symmetrical decrease in the number of inhabitants in another. This is obviously a simplification of reality, but it allows us to have decoupled equations.

The population evolution depends on two noises: $\eta_i(t)$, which describes out-of-system growth and $\zeta_i(t)$ the Lévy noise which describes interurban migrations. We characterized the distribution of both these noises and we can also estimate some of their correlations. The data doesn't allow us to test temporal correlations, as we don't have enough dates, and we will assume the standard assumption of uncorrelated noises $\langle \eta_i(t)\eta_j(t')\rangle \propto \delta(t - t')$ and $\langle \zeta_i(t)\zeta_j(t')\rangle = 0$ for $t \neq t'$.

However, the available data allows us to check for correlations between cities (at equal times) for France and the US. We then compute the correlation coefficient ρ defined as

$$\rho = \frac{\langle \eta_i\eta_j\rangle_{i\neq j} - \langle \eta_i\rangle\langle \eta_j\rangle}{\langle \eta_i\rangle\langle \eta_j\rangle} = \frac{\sum_{i\neq j}(\eta_i\eta_j)/N - (\sum_i \eta_i/N)^2}{(\sum_i \eta_i/N)^2}, \tag{9.15}$$

where N is the number of cities. For France and the US, we find for the correlation coefficients for η and ζ (denoted by ρ_η and ρ_ζ respectively), the following values:

- France: $\rho_\eta \simeq 1/100$ and $\rho_\zeta \simeq 1/100$;
- US: $\rho_\eta \simeq 5/1000$ and $\rho_\zeta \simeq 6/100$.

These results indicate very low correlations and suggest that—at least for a first approach—we can neglect correlations for both noises.

Eq. 9.14 is the city growth equation, which governs the dynamics of large urban populations. The two noises are uncorrelated and multiplicative, and the Itô convention seems more appropriate here than the Stratonovich convention because the populations at time t are computed independently of the interurban migration terms at time $t + dt$ (van Kampen, 1981). The estimates of the different parameters are summarized in Table 9.4.

To summarize our approach, the central limit theorem, given the width of the distribution of migration flows, allows us to show that many details in Eq. 9.4 are unnecessary and that the dynamics can be described by the more universal Eq. 9.14.

While the importance of migration has been noted in the past, even in the form of a stochastic differential equation (Bettencourt and Zünd, 2020), we show that a Gaussian noise is empirically unrepresentative and less appropriate than a multiplicative, zero-mean Lévy noise. This is an important theoretical change, which is quantitative but also—in the sense of complex system theory—qualitative. Indeed, the changing shape of interurban migration fluctuations also changes the way cities evolve through time.

As noted in Chapter 8, when summing up broadly distributed random variables, like migration shocks, rare but large events dominate the behavior of the sum of the random variables. Indeed, it is possible that $X_{N+1} \sim S_N$, in sharp contrast with finite variance variables where $X_{N+1} \ll S_N$. The urban translation of this last comment is that one single interurban migration wave can change the fate of a city: one migration wave can be of the typical size of the city population. The following chapter will elaborate on the consequences of this result and will propose a heuristic solution for Eq. 9.14.

10
About city dynamics

We saw in the previous chapter that starting from elementary and undisputed sources of urban growth (natural growth and migrations), we end up with a general balance equation for the dynamics of urban population:

$$\frac{\partial S_i}{\partial t} = \eta_i S_i + \sum_{j \in N(i)} (J_{j \to i} - J_{i \to j}), \qquad (10.1)$$

where η_i is a Gaussian white noise and $J_{i \to j}$ is the flow of migrations from city i to city j during time dt.

This general equation is, however, not tractable without further empirical investigations. In the previous chapter, we saw that in- and out-migrations balance *on average*—that is, $\mathbb{E}\left(\sum_{j \in N(i)} (J_{j \to i} - J_{i \to j})\right) = 0$ where $\mathbb{E}(\dots)$ is the average over the system of cities. Interestingly, the average flow from city i to j is the same as the average counter-flow. This indicates in particular that large cities have no prominent attraction. On average, Eq. 10.1 is similar to Gibrat's law, but, crucially, fluctuations of the migration balance are not negligible. This amounts to saying that we cannot, for empirical reasons, approximate the random variable $X_i = \sum_{j \in N(i)} (J_{j \to i} - J_{i \to j})$ by its average. The underlying reasons for this behavior are to be found in the complexity of cities: cities that are similar from a distance (because they have a similar population or GDP, for example) can be very different in terms of facilities, capacity for innovation, and therefore, attractiveness. All the key economic ingredients that make cities diverse are encompassed in these fluctuations and lead to a redistribution of population among the different cities. Resorting to Gibrat's formalism would be equivalent to neglecting these differences.

Fortunately, the fluctuation size of migration flows can be mathematically well understood. From our analysis (see Chapter 9), we can show that Eq. 10.1 can empirically be "simplified" to

$$\frac{\partial S}{\partial t} = \eta S + D S^{\beta} \zeta_{\alpha, \delta}, \qquad (10.2)$$

with $\beta < 1$, D a constant and ζ an unusual Lévy white noise of parameter α and skewness δ (and zero average). The Gaussian noise η has mean r and variance σ^2. This equation is simpler in the sense that it has decoupled the cities S_i of the considered system of cities. The population dynamics of a city of size S depends *only* on S. It remains, however, a complex mathematical object whose resolution is nontrivial. Upon further simplifications and assumptions, this chapter will show how Eq. 10.2 can be solved and what tangible predictions it can make.

Solving a new kind of equation

Eq. 10.2 is a two-noise equation with five parameters $(r, \sigma, \beta, \alpha, \delta)$ that can be fitted for a given country. For simplicity and upon additional (and empirically coherent) assumptions, we can reduce this number of parameters.

Parameter reduction

Two-noise stochastic differential equations are not very common objects. In Chapter 8, we saw how stochastic differential equations with Gaussian noise are well understood. We will discuss Lévy noise equations further in the following sections.

Yet, assuming that ηS and $DS^\beta \zeta_{\alpha,\delta}$ are of the same order of magnitude, one can convince oneself that a two-noise equation with Gaussian and Lévy noises is *not far* from a Lévy noise equation. Indeed, fluctuations of the Gaussian noise are, by definition, more constricted than Lévy fluctuations so that we can approximate the Gaussian noise by its average

$$\eta S \simeq rS \tag{10.3}$$

and the growth equation becomes one-noise only

$$\frac{\partial S}{\partial t} = rS + DS^\beta \zeta_{\alpha,\delta}. \tag{10.4}$$

This approximation is graphically justified by Fig. 10.1. Comparing at time t the distribution of populations resulting from numerical simulations of Eq. 10.1 with or without approximating η by its average yields similar distributions.

For simplicity and convenience, we all also assume that the skewness of the Lévy noise δ is zero. The skewness is a measure of the asymmetry of the probability distribution. It can be empirically different from zero, but analytic resolutions of Lévy SDEs are made with zero skewness. It is hence legitimate to reduce Eq. 10.1 to a one-noise-only Lévy stochastic differential equation with three parameters (r, β, α).

Fokker–Planck equation

Stochastic differential equations with a Lévy noise are not very common, and have mostly been studied in the context of Lévy walks. Our stochastic differential equation has two noises, one being Gaussian and the other Lévy distributed, a case that was apparently not discussed in previous literature. However, we proved in the previous section, that it was reasonable to consider the corresponding one noise-equation

$$\partial_t S = rS + DS^\beta \zeta_\alpha \tag{10.5}$$

on which we will focus in the following.

In the Itô interpretation, this equation is associated with a Fokker–Planck equation given by

$$\partial_t P(S,t) = -\partial_S \left(rSP(S,t)\right) + D^\alpha \frac{\partial^\alpha}{\partial S^\alpha} \left(P(S,t)S^{\alpha\beta}\right), \tag{10.6}$$

where $\frac{\partial^\alpha}{\partial S^\alpha}$ is a fractional derivative.

To sketch the derivation of this Fokker–Planck equation, we observe that populations in Eq. 10.5 are decoupled, and we can consider a functional $R[x(t)]$ of $x(t)$

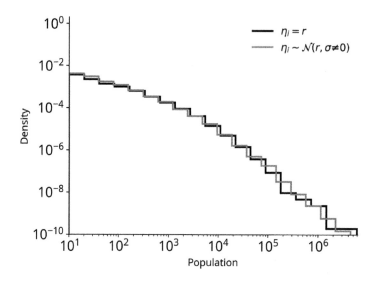

Fig. 10.1: Comparison of the average distribution population resulting from 10 numerical runs of the stochastic differential Eq. 10.1 with a Gaussian noise with finite variance σ compared with the numerical solution of Eq. 10.1, where $\eta = \langle \eta \rangle = r$ at time t. The parameters here are $\alpha = 1.3$, $\beta = 0.8$, $r = 0.01$, $\sigma = 0.06$, $D = 0.06$, $\delta = 0$ and $t = 500$.

solution of Eq. 10.5. For the sake of "notational" simplicity in this derivation, we denote here by $P(x,t) = P(x,t|x_0,0)$ the probability of having a population size x at time t (starting from x_0 at time $t = 0$). The average of this functional over this probability distribution is

$$\langle R \rangle = \int dx R(x) P(x,t), \tag{10.7}$$

which implies that

$$\frac{d}{dt} \langle R \rangle = \int dx R(x) \partial_t P(x,t). \tag{10.8}$$

On the other hand, the variation of R is

$$dR = R(x(t+dt)) - R(x(t)) \tag{10.9}$$

$$= \int dx P(x,t) \int dk R(k) \left(e^{ikx(t+dt)} - e^{ikx} \right) \tag{10.10}$$

$$= \int dx P(x,t) \int dk R(k) e^{ikx} \left(e^{ik(rxdt+kDx^\beta d\zeta)} - 1 \right), \tag{10.11}$$

where $R(k)$ denotes the Fourier transform of $R(x)$. We now use

$$\langle e^{ikDx^\beta d\zeta} \rangle = e^{-\left(Dx^\beta\right)^\alpha |k|^\alpha dt}, \qquad (10.12)$$

where the brackets indicate the average over ζ. We then expand at first order in dt and obtain

$$\langle \frac{dR}{dt} \rangle = \int dx P(x,t) \int dk R(k) e^{ikx}$$
$$\times \left[ikxr - D^\alpha x^{\alpha\beta} |k|^\alpha \right], \qquad (10.13)$$

which can be rewritten as

$$\langle \frac{dR}{dt} \rangle = \int dx P(x,t) rx \int dk R(k) e^{ikx} ik$$
$$- \int dx P(x,t) D^\alpha x^{\alpha\beta} \int dk R(k) e^{ikx} |k|^\alpha. \qquad (10.14)$$

It follows that

$$\langle \frac{dR}{dt} \rangle = \int dx P(x,t) rx \frac{dR}{dx}$$
$$- \int dx P(x,t) D^\alpha x^{\alpha\beta} \frac{d^\alpha R}{d|x|^\alpha} \qquad (10.15)$$

from which we finally obtain the Fokker–Planck equation

$$\partial_t P(x,t) = -\frac{\partial}{\partial x}(rxP(x,t)) + D^\alpha \frac{\partial^\alpha}{\partial |x|^\alpha}(x^{\alpha\beta} P(x,t)). \qquad (10.16)$$

If we rewrite this solution in terms of the population denoted by S, we obtain the Fokker–Planck equation associated with Eq. 10.5 as

$$\partial_t P(S,t) = -\partial_S (rSP(S,t)) + D^\alpha \frac{\partial^\alpha}{\partial S^\alpha}(P(S,t)S^{\alpha\beta}), \qquad (10.17)$$

where $P(S,t)$ is the probability of observing a city of population size S at time t. The quantity $\frac{\partial^\alpha}{\partial |x|^\alpha}$ is a fractional derivative, easily expressed in Fourier space.

We note here that if we had not used the approximation $\eta \simeq r$, we would have obtained the following Fokker–Planck equation

$$\partial_t P(S,t) = -\partial_S (rSP(S,t)) + \sigma^2 \frac{\partial^2 S^2 P(S,t)}{\partial S^2}$$
$$+ D^\alpha (P(S,t)S^{\alpha\beta}), \qquad (10.18)$$

where σ^2 is the variance of η. Formally, the term due to fluctuations of η will behave as k^2 in Fourier space and is thus sub-dominant (for $k \to 0$) compared with $|k|^\alpha$ ($\alpha < 2$), which corresponds to the Lévy noise.

Solving a Fokker–Planck equation with fractional derivatives

In order to illustrate the way to solve equations of the type of Eq. 10.5, we discuss here the more general case discussed by Srokowski (2009*b*) and which is of the form

$$\dot{x} = F(x) + G(x)\zeta(t), \tag{10.19}$$

where $F(x)$ is a deterministic force and ζ is an uncorrelated stochastic Lévy noise with exponent μ. In the Itô interpretation, the associated Fokker–Planck equation is

$$\frac{\partial}{\partial t}p(x,t) = -\frac{\partial}{\partial x}(F(x)p(x,t)) + \frac{\partial^\mu}{\partial x^\mu}[G(x)^\mu p(x,t)], \tag{10.20}$$

where the Riesz–Weyl fractional operator is defined in terms of its Fourier transform,

$$\frac{\partial^\mu}{\partial x^\mu} = \mathcal{F}^{-1}(-|k|^\mu). \tag{10.21}$$

Force-free case. First, we consider the force-free case $F(x) = 0$ and a power-law diffusion coefficient $G(x) = |x|^{-\theta/\mu}$ (the initial condition is always taken as $p(x,t) = \delta(x)$). The fractional Fokker–Planck equation then becomes

$$\frac{\partial p(x,t)}{\partial t} = \frac{\partial^\mu[|x|^{-\theta}p(x,t)]}{\partial|x|^\mu}. \tag{10.22}$$

In Fourier space, this equation reads

$$\frac{\partial p(k,t)}{\partial t} = -|k|^\mu \mathcal{F}[|x|^{-\theta}p(x,t)]. \tag{10.23}$$

Eq. 10.22 can be solved (Srokowski and Kamińska, 2006; Srokowski, 2009*b*) in the limit of large x using the Fox functions formalism (Fox, 1961). When $\theta = 0$, the Fourier transform of the exact solution is

$$p(k,t) = e^{-|k|^\mu t} \tag{10.24}$$

and going back to x space we obtain

$$p(x,t) \propto \int dk e^{ikx - |k|^\mu t}. \tag{10.25}$$

This function can naturally be written in terms of the Fox function $H_{2,2}^{1,1}$

$$p(x,t) = \frac{1}{\mu|x|} H_{2,2}^{1,1}\left[\frac{|x|}{(K^\mu t)^{1/\mu}} \,\middle|\, \begin{matrix} (1,1/\mu),(1,1/2) \\ (1,1),(1,1/2) \end{matrix}\right]. \tag{10.26}$$

In this simpler case, we understand how Fox functions appear in this problem: the inverse Fourier transform of a stretched exponential (of the form Eq. 10.24) leads naturally to $H_{2,2}^{1,1}$.

The Fox function (see (Fox, 1961) and (Srokowski and Kamińska, 2006) and references therein) is defined as an inverse Mellin transform in the following way:

$$
H_{pq}^{mn} \left[z \, \middle| \, \begin{matrix} (a_1, A_1), (a_2, A_2), \dots, (a_p, A_p) \\ (b_1, B_1), (b_2, B_2), \dots, (b_q, B_q) \end{matrix} \right] = \frac{1}{2\pi i} \int_L \chi(s) z^s ds, \qquad (10.27)
$$

where

$$
\chi(s) = \frac{\prod_1^m \Gamma(b_j - B_j s) \prod_1^n \Gamma(1 - a_j + A_j s)}{\prod_{m+1}^q \Gamma(1 - b_j + B_j s) \prod_{n+1}^p \Gamma(a_j - A_j s)}. \qquad (10.28)
$$

The coefficients A_i and B_i are positive and the contour L is a straight line parallel to the imaginary axis that separates the poles of both gamma functions in χ.

The Fox functions have a high number of properties, and we list here only a few of them. First, there is some sort of duality $z \leftrightarrow 1/z$:

$$
H_{pq}^{mn} \left[z \, \middle| \, \begin{matrix} (a_p, A_p) \\ (b_q, B_q) \end{matrix} \right] = H_{pq}^{mn} \left[\frac{1}{z} \, \middle| \, \begin{matrix} (1 - b_q, B_q) \\ (1 - a_p, A_p) \end{matrix} \right]. \qquad (10.29)
$$

Second, there is also a sort of multiplication rule that reads

$$
z^\sigma H_{pq}^{mn} \left[z \, \middle| \, \begin{matrix} (a_p, A_p) \\ (b_q, B_q) \end{matrix} \right] = H_{pq}^{mn} \left[z \, \middle| \, \begin{matrix} (a_p + \sigma A_p, A_p) \\ (b_q + \sigma B_q, B_q) \end{matrix} \right]. \qquad (10.30)
$$

Finally, the Fourier (cosine) transform of a Fox function is also a Fox function but with different parameters:

$$
\int_0^\infty H_{pq}^{mn} \left[x \, \middle| \, \begin{matrix} (a_p, A_p) \\ (b_q, B_q) \end{matrix} \right] \cos(kx) dx = \frac{\pi}{k} H_{q+1,p+2}^{n+1,m} \left[k \, \middle| \, \begin{matrix} (1 - b_q, B_q), (1, 1/2) \\ (1, 1), (1 - a_p, A_p), (1, 1/2) \end{matrix} \right]. \qquad (10.31)
$$

Going back to Eq. 10.22 with $\theta \neq 0$, as the solution for $\theta = 0$ can be expressed as a Fox function, it is then natural to look for a function of the form

$$
p(x, t) = N a H_{2,2}^{1,1} \left[a|x| \, \middle| \, \begin{matrix} (a_1, A_1), (a_2, A_2) \\ (b_1, B_1), (b_2, B_2) \end{matrix} \right], \qquad (10.32)
$$

where $a = a(t)$ and N is the normalization constant. The parameters a_i, A_i, b_i and B_i, and the function $a(t)$ can then be determined by comparing the expansion (for $k \to 0$) for both sides of Eq. 10.23. It is important to note here that this is possible thanks to the multiplication rule Eq. 10.30, which transforms $|x|^{-\theta} p(x)$ to a Fox function (with other parameters). Details of this calculation can be found in (Srokowski and Kamińska, 2006), and the solution of Eq. 10.22 can be written as (for large x)

$$p(x,t) = NaH_{2,2}^{1,1}\left[a|x| \, \middle| \, \begin{array}{c} (1-\frac{1-\theta}{\mu+\theta}, \frac{1}{\mu+\theta}), (1-\frac{1-\theta}{2+\theta}, \frac{1}{2+\theta}) \\ (b_1, B_1), (1-\frac{1-\theta}{2+\theta}, \frac{1}{2+\theta}) \end{array}\right].$$
(10.33)

The large asymptotic behavior is

$$p(x,t) \sim \frac{t^{\mu/\mu+\theta}}{|x|^{1+\mu}},$$
(10.34)

and in this Itô prescription, the tail of the distribution is the same as the Lévy noise.

Note that in the Stratonovich interpretation, introducing a new variable y given by

$$y(x) = \frac{\mu}{\mu+\theta}|x|^{(\mu+\theta)/\mu}\text{sgn}(x),$$
(10.35)

leads to a Langevin equation with additive noise. The corresponding Fokker-Planck equation then reads (we now denote by p_S the solution for $p(x,t)$ in this Stratonovich interpretation):

$$\frac{\partial}{\partial t}p_S(y,t) = \frac{\partial^\mu}{\partial|y|^\mu}p_S(y,t),$$
(10.36)

and can be solved exactly (in contrast with the Itô case) with a solution expressed in the form (West *et al.*, 1997*a*; Metzler and Klafter, 2000)

$$p_S(y,t) = \frac{1}{\mu|y|}H_{2,2}^{1,1}\left[\frac{|y|}{t^{1/\mu}} \, \middle| \, \begin{array}{c} (1,1/\mu), (1,1/2) \\ (1,1), (1,1/2) \end{array}\right].$$
(10.37)

Transforming back to the original variable x then leads to

$$p_S(x,t) = \frac{\mu+\theta}{\mu^2|x|}H_{2,2}^{1,1}\left[\frac{|x|^{1+\theta/\mu}}{(1+\theta/\mu)t^{1/\mu}} \, \middle| \, \begin{array}{c} (1,1/\mu), (1,1/2) \\ (1,1), (1,1/2) \end{array}\right].$$
(10.38)

In this Stratonovich case, the asymptotic behavior is

$$p_S(x,t) \sim \frac{t^{\mu/\mu+\theta}}{|x|^{1+\mu+\theta}},$$
(10.39)

which is very different from what happens in the Itô case, where the tail is governed by the noise only (and decays as $1/|x|^{1+\mu}$). Here, we observe that the tail decays with an exponent $1+\mu+\theta$ and therefore involves θ. This difference highlights the importance of the Itô-Stratonovich discussion when defining a model.

Linear force.   When there is a linear force of the form $F(x) = -\lambda x$, the Fokker–Planck equation reads (in the Itô interpretation)

$$\frac{\partial}{\partial t}p(x,t) = \lambda\frac{\partial}{\partial x}[xp(x,t)] + \frac{\partial^\mu}{\partial|x|^\mu}[|x|^{-\theta}p(x,t)], \qquad (10.40)$$

and its Fourier transform is

$$\frac{\partial}{\partial t}p(k,t) = -\lambda k\frac{\partial}{\partial k}p(k,t) - |k|^\mu \mathcal{F}_c[|x|^{-\theta}p(x,t)]. \qquad (10.41)$$

The solution of this equation can be obtained in a similar way as in the force-free case (Srokowski, 2009a). The idea is to look for a function of the form Eq. 10.32, inject it in Eq. 10.41 and expand both sides of the equation for $k \to 0$ (neglecting terms of order $|k|^{2\mu+\theta}$ and higher). In particular, the comparison of the terms of order $|k|^\mu$ leads to a differential equation for the function $a(t)$ whose solution is

$$a(t) = \left[\frac{\lambda/c_L}{1 - \exp[-\lambda(\mu + \theta)t]}\right]^{1/(\mu+\theta)}, \qquad (10.42)$$

where c_L is a complicated function of N, μ, θ, and so on. (Srokowski, 2009b). The expansion for large $|x|$ of Eq. 10.32 with these parameters gives

$$p(x,t) = N\frac{\mu+\theta}{|x|}\sum_{i=0}^{\infty}C_i(a(t)|x|)^{\theta-(\mu+\theta)i} \qquad (10.43)$$

where C_i is a complicated coefficient that depends on i, μ, θ, a_2 and A_2. This expansion is the one used in the following.

In this case also, the same equation with the Stratonovich prescription can be solved exactly, and the solution reads

$$p_S(x,t) = \frac{\mu+\theta}{\mu^2|x|}H_{2,2}^{1,1}\left[\frac{|x|^{1+\theta/\mu}}{(1+\theta/\mu)K\sigma^{1/\mu}}\,\middle|\,\begin{matrix}(1,1/\mu),(1,1/2)\\(1,1),(1,1/2)\end{matrix}\right]. \qquad (10.44)$$

Large S expansion

The solution of the Fokker–Planck equation 10.17 can be expressed in the form of a series expansion in particular cases. The analytical solution at any time can be expressed in terms of Fox H-functions seen above. The Fourier transform of a Fox function is also a Fox function, which makes them suited for solving partial derivative equations. By looking for a solution of the form of a Fox H-function in Eq. 10.17, we can write the asymptotic expansion of the solution in powers of S^{-1}. The solution can basically be found from Eq. 10.43 with the replacements $\lambda \to -r$, $\mu \to \alpha$ and $\theta \to \alpha\beta$, and the solution reads

$$P(S,t) = N\,\alpha(1-\beta)\sum_{k=1}^{\infty}C_k\frac{a(t)^{-\alpha\beta-\alpha(1-\beta)k}}{|S|^{1+\alpha\beta+\alpha(1-\beta)k}}, \qquad (10.45)$$

where

$$C_k = \frac{(-1)^k \Gamma \left(1 + \alpha(1 - \beta)k\right)}{\Gamma \left(a_2 + A_2 \left(1 + \alpha\beta + \alpha(1 - \beta)k\right)\right) \Gamma(-\frac{\alpha - \alpha\beta}{2 - \alpha\beta})k!}. \tag{10.46}$$

The quantities N, a_2 and A_2 are adjustable normalization constants:

- $a_2 = 1/2 - \alpha\beta(1 - \alpha\beta)/(2 - \alpha\beta)$
- $A_2 = 1 - 1/(2 - \alpha\beta)$
- $N = |\Gamma[\alpha\beta/(2 - \alpha\beta)]\Gamma[a_2 + A_2]/2\Gamma(1 - \alpha\beta)\Gamma[\alpha\beta/(\mu - \alpha\beta)]|$,

and the time-dependent factor

$$a(t) = \left(\frac{r/c_L}{\exp(r\alpha(1 - \beta)t) - 1}\right)^{1/\alpha(1 - \beta)} \tag{10.47}$$

with $c_L \propto D^\alpha$.

Analysis and scaling of the solution

Power-law regime

The expansion of Eq. 10.45 shows that the probability distribution of city sizes is dominated at large S by the order $k = 1$ and converges at equilibrium to a Pareto distribution of exponent $\alpha \neq 1$, apparently consistent with Zipf's law. The speed of convergence toward this power-law can be estimated by the ratio $\lambda(S, t)$ of the first and second term of the development

$$\lambda(S, t) \propto a(t)^{-\alpha(1-\beta)} S^{-\alpha(1-\beta)} \sim \frac{r}{D^\alpha} \left(\frac{e^{rt}}{S}\right)^{\alpha(1-\beta)}, \tag{10.48}$$

equilibrium being reached if $\lambda(S) \ll 1$. For values of β close to 1 (around 0.8 in the data), we have $\alpha(1 - \beta) \simeq 0.2$. Therefore,

$$\left(\frac{e^{rt}}{S}\right)^{\alpha(1-\beta)} \ll 1 \iff S > 10^5 \overline{S(t)}, \tag{10.49}$$

where $\overline{S(t)}$ is the average size of the cities at time t. The size range (cities larger than dozens of millions of people) for which we can observe a power-law distribution of parameter α is therefore, in practice, well into the upper tail of the distribution and may not exist.

We retrieve here that there is no empirical reason to observe Zipf's law nor any other stationary distribution for urban populations. The time needed to reach a steady state in city populations is too long (or alternatively the size range on which a steady-state is expected is too large). This contradicts many economic results. Also, it challenges our own result: what proof do we have that Eq. 10.45 is a valid solution for the distribution of city population?

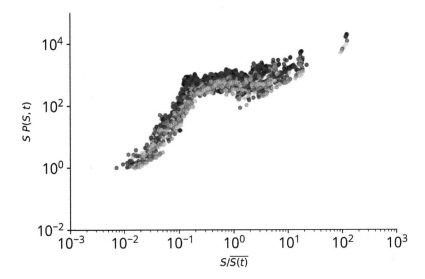

Fig. 10.2: Distribution of the quantity $P(S,t)S$ according to the ratio $S/\overline{S(t)}$ for the 500 largest French cities between 1876 and 2015. Each colour corresponds to a different year. We observe that the graphs for all years collapse into a single universal function of $S/\overline{S(t)}$, in agreement with the result of the Eq. 10.50.

Universal scaling

Eq. 10.45 gives a solution to the distribution of city population at any time t. Yet, it is not possible to assess its validity at a given time, since we know little of the initial conditions. We note, however, that from Eq. 10.45, there should be a scaling law of the form

$$P(S,t) = \frac{1}{S}F_{\alpha,\beta}\left(\frac{S}{\overline{S(t)}}\right),\tag{10.50}$$

where the function $F_{\alpha,\beta}$ depends on the country or system considered. This scaling relation is confirmed empirically for France, the only country for which the data are sufficient (see Fig. 10.2). More precisely, we observe three different regimes:

- a growing regime when $S \ll \bar{S}(t)$,
- a stable regime where $P(S,t) \sim \frac{1}{S}$ when $S \sim \bar{S}(t)$,
- a growing regime when $S \gg \bar{S}(t)$ that converges toward the relation $P(S,t) \sim \frac{1}{S^{1+\alpha}}$, as proven in Eq. 10.49.

Finally, by naively *fitting* Eq. 10.45 as if it were a power-law, we can show that the upper tail of the city size distribution can be mistaken for a Pareto distribution, with an apparent exponent, which seems to be of very good quality (R^2 very close to 1) but which varies with the definition of the upper tail, confirming a measurement artifact already mentioned in Chapter 3. When the size of the cities increases, the apparent exponent changes and can deviate considerably from 1 while converging toward the value given by α (see Fig. 10.3), as is actually observed, for example in France ($\alpha = 1.4$) and in the US ($\alpha = 1.3$).

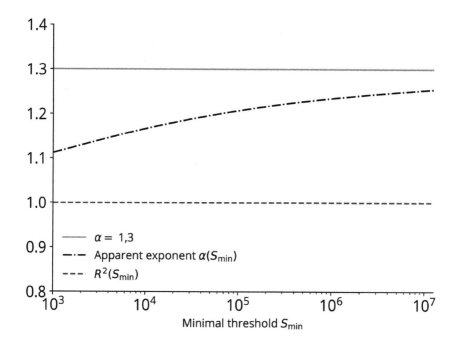

Fig. 10.3: Power-law fit of Eq. 10.45 for $\alpha = 1.3$ as a function of the minimal size S_{min} of cities. The regression of the exponent $\alpha(S_{min})$ is of very good quality $R^2 \simeq 1$), although the expansion itself is not a power-law. This illustrates the discussion in Chapter 3 that showed the danger of using fitting methods that are too simple. The apparent exponent is smaller than α but converges slowly to $\alpha = 1.3$ when the value of the threshold S_{min} increases. The parameters are here: $\alpha = 1.3$, $\beta = 0.8$, $r = 0.01$, $D = 0.06$ and $t = 500$.

Parameters over time

We considered above the solution of a general stochastic equation whose main parameters (r, α, β) are constant over time, while our estimations were made over four and five years in the four countries. Obviously, there is no reason to think *a priori* that these parameters have remained constant over decades or centuries. However, the scaling relation of Eq. 10.50 empirically verified in Fig. 10.2 (at least for France) indicates that our parameter estimates are constant enough over long periods of time.

Rank dynamics

The validity of our model can be further tested over large periods of time, in the spirit of Batty's rank clocks (see Chapter 4). Batty had shown that the dynamics of

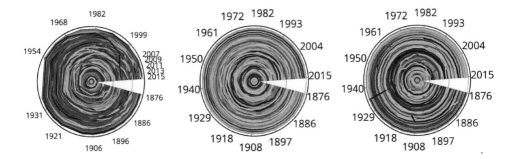

Fig. 10.4: We compare the actual dynamics of the 500 largest French cities between 1876 and 2015 (left) with Gabaix's statistical prediction (middle) and our statistical prediction (right). On the clocks, each line represents a city rank over time, where the radius is given by the rank and the angle by the time. In this representation, the largest city is at the center and the smallest at the edge of the disk.

city ranks over the long term were very turbulent and could not result from Gabaix's model (Batty, 2006). In Fig. 10.4 we compare the empirical rank clock for France from 1876 to 2015 to the results obtained with Gabaix's model and our own.

It can be seen that in Gabaix's model (middle), the rank of cities is stable on average, and not turbulent: the rank trajectories are concentric and the rank of a city oscillates around a certain level. In real dynamics, on the contrary, cities can emerge or die and very rapid changes can occur. In our model (right), the large fluctuations in the Lévy noise are able to statistically reproduce these ebbs and flows of cities, which we can show quantitatively by defining the average rank variation per unit of time

$$d = \frac{1}{NT} \sum_t \sum_{i=1}^{N} (r_i(t) - r_i(t-1)) \qquad (10.51)$$

during T years and for N cities.

The Lévy fluctuations are much better able to reproduce the turbulent dynamics of cities (see Table 10.1), which is qualitatively understandable: the foundations and disappearances of cities—due, for example, to wars, the discovery of new resources, colonization policies, and so on—are explained by migratory shocks that are too large to be compatible with the normal law. On the contrary, an essential property of Lévy stable laws (see Chapter 8) is to be able to destabilize its own sum "in one go": if the X_i are independent random variables distributed according to Lévy's law, then it is possible to find X_{n+1} so that

$$X_{n+1} \sim \sum_{i=1}^{n} X_n, \qquad (10.52)$$

which means that the fate of a particular city can change during a very short time period.

Table 10.1: The parameters of the Lévy model and the Gabaix model are fitted to the French dataset between 2003 and 2008, the US dataset between 2013 and 2017 and the UK dataset between 2012 and 2016 respectively. The most complete dataset is the French one, which contains the total population of all cities at all times, while in the US and UK only the populations of the largest cities are recorded (the top 100 in the US and the top 40 in the UK), which may explain the large discrepancies in the d value.

Average rank jump per time unit d	Data	Lévy	Gabaix
France 1876-2015	6.0	6.1	8.0
UK 1790-1990	4.7	16	27
US 1861-1991	4.8	16	25

Finally, we can compare the quality of the different models in their ability to predict the largest jump in rank of a city over time. We confirm that the Gabaix model is unable to reproduce these very large fluctuations and that our equation fits the data very well (see Fig. 10.5).

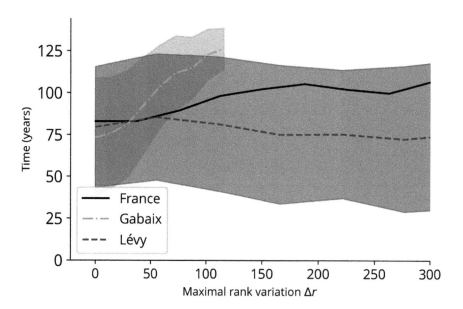

Fig. 10.5: Average number of years required to observe the maximum variation in rank Δr as a function of Δr. Although the dispersion is large, the Lévy model is compatible with the real data, unlike the Gabaix growth model.

In conclusion, we saw that microscopic details are irrelevant to the understanding of city growth and that the dynamics of cities can be represented by a universal differential equation. A crucial point in this reasoning is that, although we have on average some kind of detailed equilibrium that would lead to a Gaussian process of multiplicative growth, it is the existence of widely distributed fluctuations in inter-city migration flows that disrupt their growth to the point of changing their shape, a typical emergence phenomenon.

We also note that if a stationary regime is predicted by our model, this stationary regime is generally not reached, since in general the cities are in general out-of-equilibrium systems.

Interpretation

It is therefore the migratory shocks which govern the temporal variations of urban populations and which can, in a very short time, change the destiny of a city, the best example being that of San Francisco, which, during the Gold Rush between 1850 and 1870, grew from just over 1,000 to 150,000 inhabitants. More tragically, the city of Mariupol, Ukraine, had its population divided by five following the destruction and invasion of the city by Russia in 2022. Such extreme events are inescapable when studying cities from a historical perspective. In this book, we give them a quantitative representation, in the form of Lévy noise, which was unthinkable before.

In our opinion, this equation carries some sort of general optimistic message. In Gibrat's representation, cities grow through the accumulation of small and uncontrollable shocks. They invariably lead to a very unequal distribution of towns: Zipf's law. The underlying message is that events, however small, lead to inequality.

On the contrary, stating that one time step can change the fate of a city gives more credit to urban planning and political interventionism. Our model shows that the destiny of a city is not fixed and can be influenced by large external shocks, potentially caused by human decisions: a lot can happen in a very short time.

This equation also illustrates a very common aspect of complex systems: the emergence of qualitative shifts through "only" quantitative changes. Indeed, Eq. 10.2 is almost the same thing as Gibrat's law, except we replaced a Gaussian noise with a Lévy noise. This change is at a microscopic level, as a Lévy noise is just a generalization of a Gaussian noise with broader tails. Alternatively, if empirical migration waves had been a little smoother (with an asymptotic power-law exponent larger than 2), Gibrat's law would have been perfectly valid. This simple numerical shift, this *transition* that occurs when the power-law exponent of migration distributions goes below 2, has, however, very tangible qualitative consequences at the macroscopic level.

This is typical of emergence, and here rare but large migrations shocks can change dozens of years of city growth on very short timescales. The macroscopic result of Eq. 10.2 is hence very *different* from what would have resulted from Gibrat's law or Gabaix's model, which predict very slow shifts in city rankings in contradiction with empirical observations (Batty, 2006).

11
Outlook: Beyond Zipf's law

Zipf's law: The end?

The aim of this book is to understand the dynamics and the distribution of city populations. So far, one of the most striking empirical facts in urban studies has been Zipf's law, discovered more than a century ago. Zipf's law for cities states that the city of rank n has population S given by

$$S(n) = \frac{A}{n^\zeta},$$ (11.1)

where $\zeta \simeq 1$. Many studies have tried to explain the appearance of such a statistical regularity and never questioned its validity. With the increase of data availability, scientists started to realize that the value of ζ could significantly deviate from 1. Eventually, authors started to question the validity of the relation itself. Here, we have tried to prove that Zipf's law was more something that was hoped for than an general observation.

Ultimately the question is, however, not to understand and explain Zipf's law but rather to estimate the population distribution of cities. Starting from first principles and empirical considerations on interurban migrations, we found a distribution which is not a simple power-law. This distribution is approximately a power-law for systems comprising very large cities, and the exponent is not universal but governed by the statistics of the aforementioned migrations. Crucially, it is the existence of non-universal and broadly distributed fluctuations of the microscopic migration flows between cities that governs the statistics of city populations. This in turn shows that Zipf's law does not hold in general, but is rather a very loose approximation whose accuracy depends on various factors such as the importance of noise or finite-size effects. This result corresponds to a paradigm shift, from a statistical universality to a more complex organization of cities mainly governed by random shocks.

Given the nature of these shocks, their statistics is thus of crucial importance for understanding the population of cities and necessitate that we go beyond a simplified description in terms of power-laws. In addition, these results underline the importance of rare events in the evolution of complex systems and at a more practical level in urban planning. Indeed, not only are interurban migration flows crucial—an ingredient that is generally not considered in urban planning theories—but more importantly their large fluctuations are vital and are ultimately connected to the capacity of a city to attract a large number of new citizens.

And space?

In this book, we discussed the growth of cities in terms of their population. This is of course a very natural measure for quantifying the size of a city, but an obvious dimension remains missing in this framework: space.

Spatial structure of migration flows

Interurban migration is one of the most important factors for understanding the evolution of city populations. These migratory flows are not randomly distributed and contain some spatial information.

In (Reia *et al.*, 2022), the authors showed that intracity flows in the US usually follow a negative population density gradient, and intercity flows are concentrated in high-density core areas. Intracity flows are anisotropic and generally directed toward external counties of cities, driving asymmetrical urban sprawl. These different results can be summarized in Fig. 11.1 which shows the main migratory flows in the US.

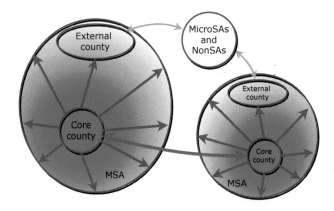

Fig. 11.1: Schematic representation of interurban migration flows obtained for the US. Figure from (Reia *et al.*, 2022).

In this figure, we observe that core counties are more likely to receive inflows from core counties of other cities than from external counties. Flows to and from micro- and non-statistical areas are more likely to be found at the external counties of a city, and intracity flows indicate vectors of redistribution of people within the city, and have an outwards radial direction: people move from central counties with larger population densities to external counties with lower densities.

These results were obtained for the US and it would be extremely interesting to confirm them for other countries. Such a result could certainly pave the way toward a better understanding of interurban migratory flows and their statistics.

Hierarchical organization of cities

Another important factor is the organization in a system of cities. Indeed, the population of a city is a very useful indicator, but its location is also crucial. More generally, the spatial distribution of cities is an important subject of debate (Ullman, 1941; Haggett *et al.*, 1977; González-Val, 2019).

Characterizing the spatial organization of urban systems is therefore a challenge which points to the more general problem of describing marked point processes in spatial statistics. In (Louail and Barthelemy, 2022), a non-parametric method was proposed that goes beyond standard tools of point pattern analysis. It is based on a mapping between the location of cities and a "dominance tree" constructed from a recursive analysis of their Voronoi tessellation. The height of a city in this tree encodes both its population and the structure of its neighborhood, and therefore reflects its importance in the system. Through the application of this tool to historical data in France and the US, several aspects of the urban dynamics can be exhibited.

However, further research is certainly needed in order to understand in detail the importance of the space and its coupling to population in the dynamics of systems of cities.

The next frontier: Spatial growth of cities

Finally, a crucial point is to understand the evolution of the spatial structure of cities and their surface area. Urban sprawl is indeed one of the most important challenges cities have to deal with to decrease their carbon footprint: it is directly connected to more car use, larger greenhouse gas emissions (Verbavatz and Barthelemy, 2019) and land artificialization.

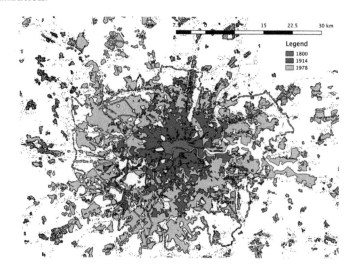

Fig. 11.2: Expansion of the city of London from 1800 to 1978. Figure courtesy of E. Strano.

The increasing availability of data (including remote sensing) allows us to observe the spatial growth of cities. An example is shown on Fig. 11.2 for the built area in London (UK). This figure immediately triggers all sorts of questions. A natural question to ask about is the temporal evolution of the surface area $A(t)$. What would be the equation for dA/dt, and how is it connected to the population evolution? Also, how is this evolution influenced by the presence of transport infrastructure? Finally, if we define the frontier of the city by some function $r(\theta, t)$ in polar coordinates, what would be the growth equation governing the evolution of $r(\theta, t)$?

Despite some interesting attempts (Zhang, 1993; Bracken and Tuckwell, 1992; Ward *et al.*, 2000; Simini and James, 2015; Carra *et al.*, 2017), it is fair to say that we did not reach the same level of understanding as for the population evolution. Such a coarse-grained quantitative description would, however, be extremely useful for identifying the critical parameters governing urban sprawl and their relative importance. This in turn could have a large impact on practical policies and the quality of life in cities.

References

Adams, John S., VanDrasek, Barbara J., and Phillips, Eric G. (1999). Metropolitan area definition in the United States. *Urban Geography*, **20**(8), 695–726.

Albert, Réka and Barabási, Albert-László (2002). Statistical mechanics of complex networks. *Reviews of Modern Physics*, **74**(1), 47.

Alves, Luiz GA, Ribeiro, Haroldo V, Lenzi, Ervin K, and Mendes, Renio S (2013). Distance to the scaling law: A useful approach for unveiling relationships between crime and urban metrics. *PLoS One*, **8**(8), e69580.

Amaral, Luís A Nunes, Buldyrev, Sergey V, Havlin, Shlomo, Leschhorn, Heiko, Maass, Philipp, Salinger, Michael A, Stanley, H Eugene, and Stanley, Michael HR (1997). Scaling behavior in economics: I. Empirical results for company growth. *Journal de Physique I*, **7**(4), 621–633.

Amir, Ariel (2020). An elementary renormalization-group approach to the generalized central limit theorem and extreme value distributions. *Journal of Statistical Mechanics: Theory and Experiment*, **2020**(1), 013214.

Arcaute, Elsa, Hatna, Erez, Ferguson, Peter, Youn, Hyejin, Johansson, Anders, and Batty, Michael (2015). Constructing cities, deconstructing scaling laws. *Journal of the Royal Society Interface*, **12**(102), 20140745.

Arribas-Bel, Daniel, Garcia-López, M-À, and Viladecans-Marsal, Elisabet (2021). Building (s and) cities: Delineating urban areas with a machine learning algorithm. *Journal of Urban Economics*, **125**, 103217.

Arshad, Sidra, Hu, Shougeng, and Ashraf, Badar Nadeem (2018). Zipf's law and city size distribution: A survey of the literature and future research agenda. *Physica A: Statistical Mechanics and its Applications*, **492**, 75–92.

Auerbach, Felix (1913). Das Gesetz der Bevölkerungskonzentration. *Petermanns Geographische Mitteilungen*, **59**, 74–76.

Barenblatt, Grigory Isaakovich (1996). *Scaling, Self-similarity, and Intermediate Asymptotics: Dimensional Analysis and Intermediate Asymptotics*. Volume 14. Cambridge University Press.

Barthelemy, Marc (2016). *The Structure and Dynamics of Cities*. Cambridge University Press.

Barthelemy, Marc (2019a). The statistical physics of cities. *Nature Reviews Physics*, **1**(6), 406–415.

Barthelemy, Marc (2019b). Tomography of scaling. *Journal of the Royal Society Interface*, **16**(160), 20190602.

Batty, Michael (2006). Rank clocks. *Nature*, **444**(7119), 592–596.

Batty, Michael (2013). *The New Science of Cities*. MIT Press.

Beckmann, Martin J (1958). City hierarchies and the distribution of city size. *Economic Development and Cultural Change*, **6**(3), 243–248.

Benguigui, Lucien and Blumenfeld-Lieberthal, Efrat (2011). The end of a paradigm: Is Zipf's law universal? *Journal of Geographical Systems*, **13**(1), 87–100.

Berry, Brian JL and Garrison, William L (1958). The functional bases of the central place hierarchy. *Economic Geography*, **34**(2), 145–154.

Bettencourt, Luis, Lobo, Jose, and Youn, Hyejin (2013). The hypothesis of urban scaling: formalization, implications and challenges. *arXiv preprint arXiv:1301.5919*.

Bettencourt, Luis and West, Geoffrey (2010). A unified theory of urban living. *Nature*, **467**(7318), 912–913.

Bettencourt, Luís MA (2013). The origins of scaling in cities. *Science*, **340**(6139), 1438–1441.

Bettencourt, Luís MA (2021). Complex networks and fundamental urban processes. In *Handbook of Cities and Networks*, pp. 41–61. Edward Elgar Publishing.

Bettencourt, Luís MA, Lobo, José, Helbing, Dirk, Kühnert, Christian, and West, Geoffrey B (2007). Growth, innovation, scaling, and the pace of life in cities. *Proceedings of the National Academy of Sciences*, **104**(17), 7301–7306.

Bettencourt, Luís M.A., Lobo, José, Strumsky, Deborah, and West, Geoffrey B. (2010). Urban scaling and its deviations: Revealing the structure of wealth, innovation and crime across cities. *PLoS One*, **5**(11), e13541.

Bettencourt, Luís M.A. and Zünd, Daniel (2020). Demography and the emergence of universal patterns in urban systems. *Nature Communications*, **11**(1), 1–9.

Bianconi, Ginestra and Barabási, Albert-László (2001). Bose-Einstein condensation in complex networks. *Physical Review Letters*, **86**(24), 5632.

Blank, Aharon and Solomon, Sorin (2000). Power laws in cities population, financial markets and internet sites (scaling in systems with a variable number of components). *Physica A: Statistical Mechanics and Its Applications*, **287**(1-2), 279–288.

Blumm, Nicholas, Ghoshal, Gourab, Forró, Zalán, Schich, Maximilian, Bianconi, Ginestra, Bouchaud, Jean-Philippe, and Barabási, Albert-László (2012). Dynamics of ranking processes in complex systems. *Physical Review Letters*, **109**(12), 128701.

Bosker, Maarten, Park, Jane, and Roberts, Mark (2021). Definition matters. Metropolitan areas and agglomeration economies in a large-developing country. *Journal of Urban Economics*, **125**, 103275.

Bouchaud, Jean-Philippe and Mézard, Marc (2000). Wealth condensation in a simple model of economy. *Physica A: Statistical Mechanics and its Applications*, **282**(3), 536–545.

Bouchaud, Jean-Philippe and Mézard, Marc (2016). *Complex systems*. École polytechnique, Palaiseau, France.

Bouchaud, Jean-Philippe and Potters, Marc (2003). *Theory of Financial Risk and Derivative Pricing: From Statistical Physics to Risk Management*. Cambridge University Press.

Bracken, Anthony J and Tuckwell, Henry C (1992). Simple mathematical models for urban growth. *Proceedings of the Royal Society of London. Series A: Mathematical and Physical Sciences*, **438**(1902), 171–181.

Bretagnolle, Anne, Pumain, Denise, and Vacchiani-Marcuzzo, Céline (2009). The organization of urban systems. In *Complexity perspectives in innovation and social change*, pp. 197–220. Springer.

Broido, Anna D and Clauset, Aaron (2019). Scale-free networks are rare. *Nature communications*, **10**(1), 1017.

Caminha, Carlos, Furtado, Vasco, Pequeno, Tarcisio HC, Ponte, Caio, Melo, Hygor PM, Oliveira, Erneson A, and Andrade Jr, José S (2017). Human mobility in large cities as a proxy for crime. *PLoS One*, **12**(2), e0171609.

Cao, Wenpu, Dong, Lei, Wu, Lun, and Liu, Yu (2020). Quantifying urban areas with multi-source data based on percolation theory. *Remote Sensing of Environment*, **241**, 111730.

Carra, Giulia, Mallick, Kirone, and Barthelemy, Marc (2017). Coalescing colony model: Mean-field, scaling, and geometry. *Physical Review E*, **96**(6), 062316.

Carroll, Glenn R (1982). National city-size distributions: What do we know after 67 years of research? *Progress in Human Geography*, **6**(1), 1–43.

Champernowne, David G (1953). A model of income distribution. *The Economic Journal*, **63**(250), 318–351.

Clauset, Aaron, Shalizi, Cosma Rohilla, and Newman, Mark EJ (2009). Power-law distributions in empirical data. *SIAM review*, **51**(4), 661–703.

Corominas-Murtra, Bernat and Solé, Ricard V. (2010). Universality of Zipf's law. *Physical Review E*, **82**(1), 011102.

Cottineau, Clémentine (2017). MetaZipf. A dynamic meta-analysis of city size distributions. *PLoS One*, **12**(8), e0183919.

Cottineau, Clémentine, Hatna, Erez, Arcaute, Elsa, and Batty, Michael (2017). Diverse cities or the systematic paradox of urban scaling laws. *Computers, Environment and Urban Systems*, **63**, 80–94.

Cournot, Antoine-Augustin (1847). *De l'origine et des limites de la correspondance entre l'algèbre et la géométrie*. L. Hachette.

de Bellefon, Marie-Pierre, Combes, Pierre-Philippe, Duranton, Gilles, Gobillon, Laurent, and Gorin, Clément (2021). Delineating urban areas using building density. *Journal of Urban Economics*, **125**, 103226.

De Gennes, Pierre-Gilles (1979). *Scaling Concepts in Polymer Physics*. Cornell University Press.

Depersin, Jules and Barthelemy, Marc (2018). From global scaling to the dynamics of individual cities. *Proceedings of the National Academy of Sciences*, **115**(10), 2317–2322.

Desrosières, Alain (2016). *La politique des grands nombres: Histoire de la raison statistique*. La découverte.

Dijkstra, Lewis, Florczyk, Aneta J, Freire, Sergio, Kemper, Thomas, Melchiorri, Michele, Pesaresi, Martino, and Schiavina, Marcello (2021). Applying the degree of urbanisation to the globe: A new harmonised definition reveals a different picture of global urbanisation. *Journal of Urban Economics*, **125**, 103312.

Dijkstra, Lewis and Poelman, Hugo (2012). Cities in europe. The new OECD-EC definition.

Dingel, Jonathan I., Miscio, Antonio, and Davis, Donald R. (2021). Cities, lights, and skills in developing economies. *Journal of Urban Economics*, **125**, 103174.

Dorogovtsev, Sergey N and Mendes, Jose FF (2002). Evolution of networks. *Advances in physics*, **51**(4), 1079–1187.

Duranton, Gilles (2015). A proposal to delineate metropolitan areas in Colombia. *Desarrollo y Sociedad* (75), 223–264.

Duranton, Gilles (2021). Classifying locations and delineating space: An introduction. *Journal of Urban Economics*, **125**, 103353.

Dyson, Freeman (2018). *The Key to Everything*. The New York Review of Book.

Eisler, Zoltán, Bartos, Imre, and Kertész, János (2008). Fluctuation scaling in complex systems: Taylor's law and beyond. *Advances in Physics*, **57**(1), 89–142.

Estoup, Jean-Baptiste (1916). *Gammes sténographiques: méthode et exercices pour l'acquisition de la vitesse*. Institut sténographique.

Feller, William (1957). *Introduction to Probability Theory and Its Applications. Vol. I.[An]*. John Wiley and Sons, Inc., Chapman and Hall.

Fermi, Enrico (1949). On the origin of the cosmic radiation. *Physical Review*, **75**(8), 1169.

Fox, Charles (1961). The G and H functions as symmetrical fourier kernels. *Transactions of the American Mathematical Society*, **98**(3), 395–429.

Fragkias, Michail, Lobo, José, Strumsky, Deborah, and Seto, Karen C (2013). Does size matter? Scaling of CO2 emissions and US urban areas. *PLoS One*, **8**(6), e64727.

Fuller, Richard A and Gaston, Kevin J (2009). The scaling of green space coverage in european cities. *Biology letters*, **5**(3), 352–355.

Gabaix, Xavier (1999). Zipf's law for cities: An explanation. *The Quarterly Journal of Economics*, **114**(3), 739–767.

Gabaix, Xavier (2009). Power laws in economics and finance. *Annu. Rev. Econ.*, **1**(1), 255–294.

Gan, Li, Li, Dong, and Song, Shunfeng (2006). Is the Zipf law spurious in explaining city-size distributions? *Economics Letters*, **92**(2), 256–262.

Gardiner, CW (1985). *Handbook of Stochastic Methods*. springer Berlin.

Gibrat, Robert (1931). *Les Inégalités Économiques*. Recueil Sirey.

Gnedenko, BV and Kolmogorov, AN (1954). *Limit Distributions for Sums of Independent Random Variables*. Addison-Wesley, Cambridge, USA.

Goldenfeld, Nigel (1992). *Lectures on Phase Transitions and the Renormalization Group*. Addison-Wesley, Advanced Book Program.

González-Val, Rafael (2019). The spatial distribution of US cities. *Cities*, **91**, 157–164.

Haag, Günter, Munz, M, Pumain, D, Sanders, L, and Saint-Julien, Th (1992). Interurban migration and the dynamics of a system of cities: 1. The stochastic framework with an application to the french urban system. *Environment and Planning A*, **24**(2), 181–198.

Haggett, Peter, Cliff, Andrew David, and Frey, Allan (1977). Locational analysis in human geography. *Tijdschrift Voor Economische En Sociale Geografie*, **68**(6).

Hairer, Martin and Pardoux, Étienne (2015). A wong-zakai theorem for stochastic pdes. *Journal of the Mathematical Society of Japan*, **67**(4), 1551–1604.

Haran, EGP and Vining Jr, Daniel R (1973a). A modified Yule-Simon model allowing for intercity migration and accounting for the observed form of the size distribution of cities. *Journal of Regional Science*, **13**(3), 421–437.

Haran, EGP and Vining Jr, Daniel R (1973*b*). On the implications of a stationary urban population for the size distribution of cities. *Geographical Analysis*, **5**(4), 296–308.

Harrison, J Michael and Williams, Ruth J (1987). Multidimensional reflected brownian motions having exponential stationary distributions. *The Annals of Probability*, 115–137.

Hill, Bruce M (1974). The rank-frequency form of Zipf's law. *Journal of the American Statistical Association*, **69**(348), 1017–1026.

Hill, Bruce M (1975). A simple general approach to inference about the tail of a distribution. *The Annals of Statistics*, 1163–1174.

Hurtado, Pablo I. (2023). Computational methods in nonlinear physics. Course on Computational Methods in Nonlinear Physics within the Master on Physics and Mathematics (FisyMat) of University of Granada.

Ijiri, Yuji and Simon, Herbert A (1975). Some distributions associated with Bose-Einstein statistics. *Proceedings of the National Academy of Sciences*, **72**(5), 1654–1657.

Iñiguez, Gerardo, Pineda, Carlos, Gershenson, Carlos, and Barabási, Albert-László (2022). Dynamics of ranking. *Nature Communications*, **13**(1), 1–7.

Ioannides, Yannis M and Overman, Henry G (2003). Zipf's law for cities: An empirical examination. *Regional Science and Urban Economics*, **33**(2), 127–137.

Kleiber, Max (1947). Body size and metabolic rate. *Physiological Reviews*, **27**(4), 511–541.

Kleiber, Max et al. (1932). Body size and metabolism. *Hilgardia*, **6**(11), 315–353.

Krugman, Paul (1996). Confronting the mystery of urban hierarchy. *Journal of the Japanese and International Economies*, **10**(4), 399–418.

Kühnert, Christian, Helbing, Dirk, and West, Geoffrey B (2006). Scaling laws in urban supply networks. *Physica A: Statistical Mechanics and Its Applications*, **363**(1), 96–103.

Kumamoto, Shin-Ichiro and Kamihigashi, Takashi (2018). Power laws in stochastic processes for social phenomena: An introductory review. *Frontiers in Physics*, **6**, 20.

Leitao, JC, Miotto, JM, Gerlach, M, and Altmann, EG (2016). Is this scaling nonlinear? *arXiv preprint arXiv:1604.02872*.

Levy, Moshe and Solomon, Sorin (1996). Power laws are logarithmic boltzmann laws. *International Journal of Modern Physics C*, **7**(04), 595–601.

Lévy, Paul (1926). Calcul des probabilités. *Revue de Métaphysique et de Morale*, **33**(3).

Lobo, José, Bettencourt, Luís MA, Strumsky, Deborah, and West, Geoffrey B (2013). Urban scaling and the production function for cities. *PLoS One*, **8**(3), e58407.

Louail, Thomas and Barthelemy, Marc (2022). A dominance tree approach to systems of cities. *Computers, Environment and Urban Systems*, **97**, 101856.

Louf, Rémi (2015). *Wandering in cities: A statistical physics approach to urban theory*. Ph.D. thesis, Université Pierre et Marie Curie, Paris, France.

Louf, Rémi and Barthelemy, Marc (2014*a*). How congestion shapes cities: from mobility patterns to scaling. *Scientific Reports*, **4**(1), 5561.

Louf, Rémi and Barthelemy, Marc (2014*b*). Scaling: lost in the smog. *Environment and Planning B: Planning and Design*, **41**, 767–769.

Mandelbrot, Benoit (1960). The pareto-levy law and the distribution of income. *International economic review*, **1**(2), 79–106.

Mandelbrot, Benoît and Hudson, Richard L. (2007). *The Misbehavior of Markets: A Fractal View of Financial Turbulence*. Basic Books.

Mantegna, Rosario N and Stanley, H Eugene (1995). Scaling behaviour in the dynamics of an economic index. *Nature*, **376**(6535), 46–49.

Marsili, Matteo and Zhang, Yi-Cheng (1998). Interacting individuals leading to Zipf's law. *Physical Review Letters*, **80**(12), 2741.

Martínez-Mekler, Gustavo, Martínez, Roberto Alvarez, del Río, Manuel Beltrán, Mansilla, Ricardo, Miramontes, Pedro, and Cocho, Germinal (2009). Universality of rank-ordering distributions in the arts and sciences. *PLoS One*, **4**(3), e4791.

Massey Jr, Frank J (1951). The Kolmogorov-Smirnov test for goodness of fit. *Journal of the American Statistical Association*, **46**(253), 68–78.

Metzler, Ralf and Klafter, Joseph (2000). The random walk's guide to anomalous diffusion: a fractional dynamics approach. *Physics Reports*, **339**(1), 1–77.

Molinero, Carlos and Thurner, Stefan (2019). How the geometry of cities explains urban scaling laws and determines their exponents. *arXiv preprint arXiv:1908.07470*.

Moran, José (2020). *Statistical physics and anomalous macroeconomic fluctuations*. Ph.D. thesis, EHESS, Paris.

Moreno-Monroy, Ana I., Schiavina, Marcello, and Veneri, Paolo (2021). Metropolitan areas in the world. Delineation and population trends. *Journal of Urban Economics*, **125**, 103242.

NASA (2022). Earth data website.

Nomaler, Önder, Frenken, Koen, and Heimeriks, Gaston (2014). On scaling of scientific knowledge production in us metropolitan areas. *PLoS One*, **9**(10), e110805.

Oliveira, Erneson A, Andrade Jr, José S, and Makse, Hernán A (2014). Large cities are less green. *Scientific reports*, **4**.

Pumain, Denise (2004). Scaling laws and urban systems. *Santa Fe Institute, Working Paper n 04-02*, **2**, 26.

Pumain, Denise and Moriconi-Ebrard, François (1997). City size distributions and metropolisation. *Geojournal*, **43**(4), 307–314.

Pumain, Denise, Paulus, Fabien, Vacchiani-Marcuzzo, Céline, and Lobo, José (2006). An evolutionary theory for interpreting urban scaling laws. *Cybergeo: European Journal of Geography*, **2006**, 1–20.

Reia, Sandro M, Rao, P Suresh C, Barthelemy, Marc, and Ukkusuri, Satish V (2022). Spatial structure of city population growth. *Nature Communications*, **13**(1), 5931.

Ribeiro, Fabiano L and Rybski, Diego (2021). Mathematical models to explain the origin of urban scaling laws: a synthetic review. *arXiv preprint arXiv:2111.08365*.

Rosen, Kenneth T. and Resnick, Mitchel (1980). The size distribution of cities: An examination of the Pareto law and primacy. *Journal of Urban Economics*, **8**(2), 165–186.

Rozenfeld, Hernán D, Rybski, Diego, Andrade, José S, Batty, Michael, Stanley, H Eugene, and Makse, Hernán A (2008). Laws of population growth. *Proceedings of the National Academy of Sciences*, **105**(48), 18702–18707.

Rozenfeld, Hernán D, Rybski, Diego, Gabaix, Xavier, and Makse, Hernán A (2011). The area and population of cities: New insights from a different perspective on cities. *American Economic Review*, **101**(5), 2205–2225.

Rybski, Diego, Buldyrev, Sergey V, Havlin, Shlomo, Liljeros, Fredrik, and Makse, Hernán A (2009). Scaling laws of human interaction activity. *Proceedings of the National Academy of Sciences*, **106**(31), 12640–12645.

Rybski, Diego, Sterzel, Till, Reusser, Dominik E, Winz, Anna-Lena, Fichtner, Christina, and Kropp, Jürgen P (2017). Cities as nuclei of sustainability? *Environment and Planning B: Urban Analytics and City Science*, **44**(3), 425–440.

Samaniego, Horacio and Moses, Melanie E (2008). Cities as organisms: Allometric scaling of urban road networks. *Journal of Transport and Land use*, **1**(1).

Shalizi, Cosma Rohilla (2011). Scaling and hierarchy in urban economies. *arXiv preprint arXiv:1102.4101*.

Simini, Filippo and James, Charlotte (2015). Discovering the laws of urbanisation. *arXiv preprint arXiv:1512.03747*.

Simkin, Mikhail V and Roychowdhury, Vwani P (2011). Re-inventing Willis. *Physics Reports*, **502**(1), 1–35.

Simon, Herbert A. (1955). On a class of skew distribution functions. *Biometrika*, **42**(3/4), 425–440.

Singer, Hans W (1936). The "courbe des populations". A parallel to pareto's law. *The Economic Journal*, **46**(182), 254–263.

Solomon, Sorin (1998). Stochastic Lotka-Volterra systems of competing autocatalytic agents lead generically to truncated Pareto power wealth distribution, truncated Levy-stable intermittent market returns, clustered volatility, booms and crashes. In *Decision Technologies for Computational Finance: Proceedings of the Fifth International Conference Computational Finance*, pp. 73–86. Springer.

Solomon, Sorin and Richmond, Peter (2001). Power laws of wealth, market order volumes and market returns. *Physica A: Statistical Mechanics and Its Applications*, **299**(1-2), 188–197.

Solomon, Sorin and Richmond, Peter (2002). Stable power laws in variable economies; Lotka-Volterra implies Pareto-Zipf. *The European Physical Journal B-Condensed Matter and Complex Systems*, **27**, 257–261.

Soo, Kwok Tong (2005). Zipf's law for cities: A cross-country investigation. *Regional Science and Urban Economics*, **35**(3), 239–263.

Sornette, Didier and Cont, Rama (1997). Convergent multiplicative processes repelled from zero: Power laws and truncated power laws. *Journal de Physique I*, **7**(3), 431–444.

Srokowski, Tomasz (2009a). Fractional Fokker-Planck equation for Lévy flights in nonhomogeneous environments. *Physical Review E*, **79**(4), 040104.

Srokowski, Tomasz (2009b). Multiplicative Lévy processes: Itô versus Stratonovich interpretation. *Physical Review E*, **80**(5), 051113.

Srokowski, T and Kamińska, A (2006). Diffusion equations for a Markovian jumping process. *Physical Review E*, **74**(2), 021103.

Stauffer, Dietrich and Aharony, Ammon (2018). *Introduction to Percolation Theory*. Taylor & Francis.

Steindl, Josef (1965). *Random Processes and the Growth of Firms: A Study of the Pareto Law*. Volume 18. Hafner Publishing Company.

Strano, Emanuele and Sood, Vishal (2016). Rich and poor cities in Europe. an urban scaling approach to mapping the European economic transition. *PLoS One*, **11**(8), e0159465.

Taylor, Lionel Roy (1961). Aggregation, variance and the mean. *Nature*, **189**(4766), 732–735.

Ullman, Edward (1941). A theory of location for cities. *American Journal of Sociology*, **46**(6), 853–864.

US Office of Management and Budget (2010). 2010 standards for delineating metropolitan and micropolitan statistical areas. Technical report, Washington D.C., United States of America.

van Kampen, Nicolaas G. (1981). Itô versus Stratonovich. *Journal of Statistical Physics*, **24**(1), 175–187.

Verbavatz, Vincent and Barthelemy, Marc (2019). Critical factors for mitigating car traffic in cities. *PLoS One*, **14**(7), e0219559.

Verbavatz, Vincent and Barthelemy, Marc (2020). The growth equation of cities. *Nature*, **587**(7834), 397–401.

Vining Jr, Daniel R (1974). On the sources of instability in the rank-size rule: Some simple tests of Gibrat's law. *Geographical Analysis*, **6**(4), 313–329.

Ward, Douglas P, Murray, Alan T, and Phinn, Stuart R (2000). A stochastically constrained cellular model of urban growth. *Computers, Environment and Urban Systems*, **24**(6), 539–558.

West, Bruce J, Grigolini, Paolo, Metzler, Ralf, and Nonnenmacher, Theo F (1997a). Fractional diffusion and Lévy stable processes. *Physical Review E*, **55**(1), 99.

West, Geoffrey B, Brown, James H, and Enquist, Brian J (1997b). A general model for the origin of allometric scaling laws in biology. *Science*, **276**(5309), 122–126.

Willis, John C and Yule, G Udny (1922). Some statistics of evolution and geographical distribution in plants and animals, and their significance. *Nature*, **109**(2728), 177–179.

Yule, George Udny (1925). II.—A mathematical theory of evolution, based on the conclusions of Dr. JC Willis, F.R.S. *Philosophical transactions of the Royal Society of London. Series B, Containing Papers of a Biological Character*, **213**(402-410), 21–87.

Zanette, Damián H and Manrubia, Susanna C (1997). Role of intermittency in urban development: A model of large-scale city formation. *Physical Review Letters*, **79**(3), 523.

Zhang, Wei-Bin (1993). An urban pattern dynamics with capital and knowledge accumulation. *Environment and Planning A*, **25**(3), 357–370.

Zipf, George Kingsley (1949). *Human Behavior and the Principle of Least Effort*. Addison–Wesley Press.

Index

A

Absolute Konzentration 39
Administrative boundaries 3
Asymptotic expansion 131
Auerbach 39

B

Balance equation 112
Barabási–Albert model 83
Bienaymé–Galton–Watson 84
Bose–Einstein statistics 83
Bouchaud–Mézard model 96
Branching processes 84
Broad distributions 36
Brownian motion 67

C

Cauchy distribution 109, 110
Central limit theorem 103
Champernowne's model 90
Characteristic function 104
City Clustering Algorithm (CCA) .. 9
City definition 3, 12
Colored noise 72
Core Based Statistical Area (CBSA) 8
Correlations 122

D

Detailed balance 118
Diffusion with noise 94, 96

F

Fitting power-laws 17, 42
Fokker–Planck equation 57, 73, 75, 91, 125
Fox function 128, 129
Fractional derivative 125, 128
Friction 90

F

Functional definition 5
Functional Urban Area (FUA) ... 4, 6

G

Gabaix's model 92
Galton–Watson process 84
Generalized central limit theorem 105, 110, 120
Generalized Lotka–Volterra equation 98
Geometric Brownian motion 91
Giant component 9
Gibrat's law 77, 88
Gravitational model 117
Gridded Population of the World (GPW) 10
Growth equation of cities 120, 124

H

Haran–Vining model 76, 94
Hierarchical organization of cities . 35, 39, 85, 140
Hill estimator 48
Hill plot 48

I

Interurban migrations ... 95, 115, 118
Itô vs. Stratonovich 69
Itô's prescription 70, 71

K

Kolmogorov–Smirnov estimate 48
Kolmogorov–Smirnov statistics 48

L

Lévy distribution 110
Lévy stable laws 105, 108
Langevin equation 69
Langevin noise 69

150

Larger Urban Zone (LUZ) 7
Law of large numbers 103
Lorentzian distribution 110

M

Markovian process 68
Master equation 95
Maximum likelihood estimator (MLE) 44
Megacities 14
Metropolitan Statistical Area (MSA) 4, 7
Migration flows: Spatial structure 139
Migration graph 115
Migrations 94, 118, 137
MLE for power-laws 46
Models of urban growth 76
Morphological definition 5
Multiplicative growth 76
Mystery of urban hierarchy 35

N

Noise correlations 122
Noise-induced transition 55
Normal distribution 109

O

Occam's razor .... ii, 19, 93
Optimal size of cities 35
Ordinary-least-square regression (OLS) 42
Ornstein–Uhlenbeck noise 72
Out-of-system growth 113

P

Pareto 36, 37
Paris (France) 4
Path dependency 17
Percolation 9
Population distribution 35
Power-laws 17, 35, 36, 46
Preferential attachment 76
Primacy index 39

R

Random proportional growth 77

Random walk 67
Rank clocks 63, 134
Rank dynamics 55, 59, 61, 134
Rank flux 59
Rank stable 57
Rank turnover 59
Rank variations 63
Ranking 54
Reaction-diffusion model 87
Real-space renormalization approach 105
Riez–Weyl operator 128

S

Scaling 14, 31
Scaling laws 16
Score stable 57
Simon's model 77, 79, 85
Skewness 108
Social phenomena 35
Spatial growth of cities 140
Spurious exponents 50
Steindl's model 87
Stochastic differential equation 67, 69, 125
Stratonovich's prescription 70
Strong Gibrat's law 77
Symmetric random walk 68
Systems of cities 39

T

Tauberian theorems 108
Turnover rate 59

U

Urbanization rate 13

W

Weak Gibrat's law 77
White noise 72
Wiener process 68
Wright–Fisher equation 62

Y

Yule versus Simon 81
Yule's model 77

Z

Zhang–Marsili model 86

Zipf's exponent 51, 98

Zipf's law 38, 39, 51, 76, 138